DREADNOUGHT TO POLARIS

Dreadnought to Polaris

MARITIME STRATEGY SINCE MAHAN

Papers from the Conference on Strategic Studies at the University of Western Ontario, March 1972.

Edited by A.M.J. Hyatt

THE COPP CLARK PUBLISHING COMPANY
Toronto   Montreal   Vancouver
NAVAL INSTITUTE PRESS
Annapolis, Maryland

© The Copp Clark Publishing Company 1973

All rights reserved. No part of the material covered by this copyright may be reproduced in any form or by any means (whether electronic, mechanical or photographic) for storage in retrieval systems, tapes, discs or for making multiple copies without the written permission of The Copp Clark Publishing Company.

Printed and bound in Canada.

ISBN 0 7730 4010 2

Editor's Introduction

More than eighty years ago Admiral A.T. Mahan declared that "historians generally have been unfamiliar with the conditions of the sea, having as to it neither special interest nor special knowledge; and the profound determining influence of maritime strength upon great issues has consequently been overlooked." Things have not changed much, for in Professor Schurman's opinion "the amount of knowledge or even interest in naval history or strategy on the part of scholars at large is not great." Among historians, or for that matter "scholars at large" few have worked more diligently and successfully against this trend than Gerald S. Graham, who in 1971-72 was the first Professor of Strategic Studies at the University of Western Ontario. Professor Graham was largely responsible for the Conference on Strategic Studies at which the papers here collected were presented to an audience of teachers, students, sailors, scholars and interested individuals.

The nature of the papers varies a great deal — from the meticulous analysis of a British naval failure by Professor Marder to informed speculation on the future of maritime strategy. While most of the papers deal with strategic problems on a global scale, specifically Canadian problems also find their place. Taken together, the papers by no means present a comprehensive history of maritime strategy in the twentieth century, but they do offer a variety of insights into aspects of it. The problems of maritime strategy are approached from a variety of angles.

Unfortunately the discussions and commentaries which followed each paper cannot be reproduced, but thanks are due to Professors G.S. Graham, K.H.W. Hilborn, G. Murray of the University of Western Ontario, Professor Albert Legault of Laval and Mr. Robert Reford of the Canadian Institute of International Affairs, all of whom acted as chairmen-commentators for the various sessions. The editor is grateful to all those who presented papers for their prompt cooperation in revising the essays for publication.

The conference was jointly sponsored by the History Department of the University of Western Ontario, as part of its Strategic Studies programme, and by the Canadian Institute of International Affairs. A grant from the Canada Council, which assisted in making the conference possible is gratefully acknowledged.

A.M.J. Hyatt
University of Western Ontario

Contents

Editor's Introduction
A.M.J. Hyatt, University of Western Ontario

1 An Historian and the Sublime Aspects of the
  Naval Profession   1
  D.M. Schurman, Queen's University

2 German Seapower: A Study in Failure   12
  Theodore Ropp, Duke University

3 The R.N.A.S. in Combined Operations 1914-1915   19
  *Commander W.A.B. Douglas, Directorate of History,
  Canadian Forces Headquarters*

4 The Dardanelles Revisited: Further Thoughts on
  the Naval Prelude   30
  Arthur Marder, University of California, Irvine

5 Smaller Navies and Disarmament: Sir Herbert
  Richmon's 'Small Ship' Theories and the
  Development of British Naval Policy in the 1920's   47
  B.D. Hunt, The Royal Military College of Canada

6 Canadian Maritime Strategy in the Seventies   64
   *G.R. Lindsey, Defence Research Analysis
   Establishment, Ottawa*

7 Problems of Naval Arms Control: The High Seas,
   The Deep Ocean and the Seabed   77
   *William Epstein, Disarmament Affairs Division of
   the United Nations*

8 The Seas in the Seventies   89
   *Lieutenant-Commander A.D. Taylor, Department of
   External Affairs, Ottawa*

9 From Polaris to the Future   101
   *Ian Smart, Assistant Director, International
   Institute for Strategic Studies*

   References   111

1 An Historian and the Sublime Aspects of the Naval Profession

D. M. SCHURMAN*
Queen's University

The remarks that I wish to address to this gathering of what Mahan's Admiral Henry Erben[1] would have called quill-drivers will not be new to many of you. At the moment, I have a manuscript on the life of Sir Julian Corbett complete. When published, it will complement the approach to his historical works available in my book, *The Education of a Navy*.[2] Indeed, Corbett has been a part of my research life for so long that every time I try to say something about him in public I feel as though I am robbing one part of my mind in order to give substance to another. However, the re-use of one's own materials in different form has strong traditions of the military historian's trade to support it. Both Sir Basil Liddell Hart and Sir Herbert Richmond were strong practitioners of the realignment of their own work for the sake of emphasis. To hide behind such eminent cloaks seems as good a way as any to justify my proceeding.

This gathering is perhaps the wrong place to say it, but it seems to me that no scholar whose historical bent is in a naval direction ought to apologize too much for reiteration. The amount of knowledge or even interest in naval history or strategy on the part of scholars at large is not great. Even the British, with their national position literally founded upon the seas, have given more lip service

---

*Aside from identifying some figures who may not be well known, and aside from one or two generally available quotations, much of the material comes from my forthcoming biography of Sir Julian Corbett. The references will appear in their proper place in due course. The author reserves the copyright on all such material and on its interpretation.

than understanding to their sea-going past. A personal reference may help to clarify this. When, years ago, I was being interviewed for an award, in Cambridge, the one question for which I was totally unprepared was "Why should we subsidize such parochial work?". At that time, my obvious incredulity that the question should even be asked saved me from attempting a dilation on the national importance of the Royal Navy in fifteen minutes. I am by no means certain, modesty aside, that it was the nature of my subject that secured for me the prize I sought. Since then, looking at the attention lavished by publishers on soldier boys as opposed to that given to jack tar has kept my original suspicion that I work in an historical backwater firm. That Professor Graham has gathered together so many distinguished exceptions to my point reanimates my already considerable respect for him.

If my references are mainly to the Royal Navy, I should point out that what knowledge I have of naval strategy comes from historical study of that service and its chroniclers. No claim is here made to introduce so-called universal appreciations. What has captured my interest has been what British sailors did, how they chronicled it, and to a lesser extent, how they attempted to codify the resulting combination of chronicle and reflection. As far as relevance is concerned, one is often enough told by those modern practitioners of the astrologer's art, the political scientists, that history is merely good background study, to incline one to admit that theirs is a different kind of academic discipline, and in no sense a rival.

The first thing I wish to do is to look at British naval strategy as it developed in the Age of Sail. Between the reign of Queen Elizabeth and the latter portions of that of George III, naval strategy was a product of the combined appreciation of practical sailors, overseas administrators of Empire, and practicing statesmen. The last group were the most important for it was in the minds of statesmen, and in their codification of past experience when issuing specific instructions to naval commanders and concerning naval practice, that strategy achieved shape and form. When, therefore, a naval commander on a distant station received a communication to the effect that "My Lords think it right that you should proceed etc." then that professional sailor was being put into possession of what naval strategy there was. It is also worth noting that this strategic instruction, in the form of instruction, would represent at any time a cumulative process. Precedent was important — almost as important as the circumstance to which it was joined, more or less felicitously according to the mental power of the then incumbent men of power in London.

For instance, there is a direct link between the strategy that was almost inadvertently adopted by the wild birds that swooped on and nibbled the edges of the Armada in 1588, from to windward in the Channel, and the accepted view by 1805. One discovered and the other knew that to keep to windward of any invading force in the same Channel would prevent a seaborne foe from landing a force on the British mainland strong enough to successfully prosecute a decisive land campaign. Between the two dates lay some two hundred years of experience, and by the latter one the dimly understood, almost reflex, and perhaps lucky dispositions that had once characterized the forces of Lord Howard, were appreciated as proper contingency strategy by Captains, Admirals, First Lords, and Prime Ministers. This example is taken because, clearly, it does not bear any direct relation to abstract ideas known as principles of war for general sea as opposed to land use, but is related directly to ship type, fleet capability, circumstances, and the inescapable facts of geography. British strategy did not exist in a special codified set of rules universally applicable to all situations but to guide-rules that had been tested over time and, barring the effects of luck and misapplication, were certified to produce concrete results in particular circumstances. This strategy was pragmatic. It worked. However, the fear exhibited in Great Britain over the possibility that Napoleon actually might invade testifies to the fact that, to many people in power and out, naval strategy was regarded as a mysterious projection whose success would be bound more by luck than it would rest on a sure appreciation of the securities inherent in proper sea dispositions.

Another thing to note about Britain's sea strategy during this long period is that, although it had to do with trade and attacks on enemy possessions in war, at heart it was defensive in purpose. As the medieval chronicler wrote "Keep ye the narrow see as your twin eyne." This was a constant in British sea strategy, and it remained so until, not only 1815, but 1940.

Finally, it should be noted that while the Napoleonic wars induced their progeny, Clausewitz and Jomini, to ferret out the general rules of military land conduct — or at any rate to base their conclusions on the study of land warfare — the extended conflict that included Trafalgar in its records did not produce any such thing. Nothing, of significance, in fact emerged but the belated compilation of the writings of the dead hero. The main interest was in the elan and the attractive eager temperament of Nelson rather than in his professional competence on the grand scale. (It will be noted that I am not here discussing tactics, a subject upon which much public discussion took place and about which weighty works were written.)

It would be quite wrong to deduce from the above that defence against invasion was the sole preoccupation of statesmen who were concerned with naval strategic dispositions. On the contrary. During the great struggle against Napoleon, the defence and attack of trade and territory loomed large, both for the immediate practical purposes of the moment, and when viewed from the point of view of the peace negotiations that were bound to ensue. The sum of British naval strategy was a sophisticated process of changing values and expedients, and sometimes clearly, of changing priorities.

After 1815, the multitudinous duties that devolved upon armed sea forces did not lessen, even if the pressures upon priorities did. For a quarter of a century, or slightly more, the British were not able to discern any serious rival to their sea-control in the world. What the effect of their sea activity in the Napoleonic era actually was might be debated. What was not debated was that the victors of Trafalgar were the First Viceroys of Neptune.

However, these halcyon days of sea hegemony did not last. Into their twilight days moved a revolution in ship design — as profound for the time as any rocket dominated power alteration of the mid-twentieth century. It is very important to note that this conjunction of pride of place in Neptune's world, and the unsettling effect of ongoing ship design and materièl change brought thought about strategy, thought that involved the assignment of urgent priorities to almost a dead stop. Secure in the remembrance and exercise of supremacy, with hundreds of armed craft cresting the oceans of the world, the British reaction to the naval revolution was one of unhurried experimentation and careful search for the ultimate sea-weapon of the future.[3]

When experimentation settled down by about 1885, there was a need for strategic guidance in the use of this force. The French navy was sufficient to make the British nervous in the period 1885-1895: the rise of the German navy after the latter date induced feelings of worry that were to contain, at times, elements of panic. In short, in a world where the number of competitors for sea-supremacy was increasing rapidly what criterion should be used to the measure of the adequacy of the new steam-steel-battleship navy?

Mahan, of course, answered the call after a fashion. His work pointed to the value of simply outclassing and outnumbering the battle-wagons of the enemy or enemies. His work made an immediate appeal in England, even as it did in Germany. It is interesting and perhaps significant that this answer should have come from an *historian* (even a foreign one) rather than from a statesman. The fact is, that the old rule that statesmen made sea-strategy was not less true, but in the absence of testing naval warfare, in the face of design flux, and in the face of indecision amongst professional advisors, the politicians no longer felt the good footing of developed tradition under their feet. It was no doubt assumed that such a profound thinker as Mahan would provide proper rules for conduct from the historical logic contained in his famous case books. What the professional *sailors* wanted to do, of course, was to make sure that they had enough superior equipment to defeat a challenging force of the enemy whoever he was when war should cause a meeting. The conjunction between history and materièl came precisely at this point — the point where the materièl conscious professional planners (who now had to educate their political masters) found that they could find support for their equipment-conscious views from the pages of history. To suggest that the majority of these men held either Mahan or history in great affection would be naive. What they held in affection were their own materièl dominated views, but they were not above supporting them with any ammunition that came to hand. Their views, to be blunt, never passed the point of being convinced that they and their ships had been created for a great fleet action. Who would be the first to overshadow Nelson's exploits but on the deck of an ironclad? That was the question.

This viewpoint did not represent a sophisticated modern naval strategy. After all, it rested in the hands of men whose view of the world was too limited to risk the security of the State to their hands. (How different it had been with Lord Barham[4] and Nelson!) But Mahan and 'bash em up in a big battle' were all that the Admirals or anyone else, apparently, had to offer. Naval debate, after 1900 and Lord Fisher notwithstanding, turned on questions of materièl — a sensible enough thing had it not driven all else to the wall.

An alternative was provided, however, by a British writer, Julian Stafford Corbett. Corbett was an historian without being an academic, a feat that he managed by being financially independent. No Deans for him! His history had this difference from Mahan's — that while Mahan studied secondary sources to discover immutable rules with which to promote the intelligent growth of the American Navy, Corbett studied primary sources of British history to attempt to discover what rules for present conduct lay in the British past.

However, and this is very important, no matter how he used his historical researches to bolster his opinions on naval strategy over the next fifteen years, the fact is that his first book, *Drake and the Tudor Navy*,[5] was written simply to find out what had gone on at sea in the Elizabethan era. People like J.A. Froude wrote heroic tales that seemed to be nothing more than accounts of semi-piratical cross-ravaging. Corbett thought he discovered something much more in this activity. From a study of Drake, the Armada and the Spanish Main, he extracted the idea that the security of the coasts of England, and the best mode of attack, was guaranteed by operations off the enemy's coast — by blockade together with shore attacks, rather than by waiting to be attacked nearer to home. These views were seriously put to the Queen as an alternative to placing too much reliance on the value of army operations in the Low Countries. Sea power conferred strike mobility. As we know, and as Sir Walter Raleigh remarked, the Queen "did all by halves" and so the Elizabethan success was limited.[6] Then in his second book, *The Successors of Drake*,[7] Corbett drew the strong conclusion that for sea-power to be effective it must be able to draw on and work in concert with a strong professional army. These two concepts of strategy then, that a sea power possessed general military selective mobility and that that mobility gained strength as it was supported by a solid army partner, came out of a work originally written for historical purposes only.

At this point, he was asked to link history and strategy for lectures at the Naval War Course.[8] He then had to face the question of whether his Tudor studies held any relevance for the new iron and steam navy of the *fin de siecle*. From this point on it may be debated whether his historian's purpose dominated his strategic task or vice versa. Despite a terrific enthusiasm for what he had seen of the seamanship of the Navy, frankly Corbett was appalled when he had occasion to observe their senior brains attempting to grapple with strategic — cum-historical problems. It is true that the Naval Educational system was rudimentary, but the effect of people looking for simple effects in complicated materials struck him as almost ludicrous. Presenting carefully balanced lectures involving the interplay of intricate cause and effect he soon discovered that broadside bashing and seamanship composed the intelligent Admiral's strategic mentality. By 1906, he realized that until he could get a common body of strategic knowledge into naval officers' heads he could not hope to give lectures from which all could reasonably profit. Therefore in 1906, he and the then Course Director, Captain Edmond Slade[9], produced 'Notes on Strategy' — a strategic primer based on history and common sense. Quite properly Admiral Sir Reginald Custance[10] noted that this was the wrong way to do it — the historian should provide the history so each man could draw his own conclusions from study. Custance was right, of course, but how was one to proceed with short course senior Admirals who were both superior in manner and slow on formal thought processes?

All the long years before World War I, Corbett laboured at his job of teaching Captains and Admirals to understand what he called "the sublime aspects of their profession." He seems to have felt that he was always on the verge of success. One great wooden-headed member of his audiences was Captain Doveton Sturdee, a strong materièl man and the later victor of the Falkland Islands. Corbett wrote to his wife once — in oblique reference to Bernard Shaw's play — that "we are playing a game here — Captain Sturdee's conversion, one more try and we shall have him." Sturdee was not converted. Years later, Corbett's efforts at teaching were described, maliciously, by a former pupil. "He was always most dangerous when most sauve and when he would say 'gentlemen, mentioning a subject on which you are more knowledgeable than I' you knew he was about to craftily get across some point of his own that would be in revolt against accepted thought."[11] In short, he was too clever by half for his audiences and they resented his verbal supremacy.

The main point of strategy that he was most concerned to promote was the idea that national policy and need should govern naval activity. His audiences, materièl oriented, were convinced that all of naval strategy was ultimately bound up in the presupposition that, if you destroyed the main enemy fleet or fleets, then all other requisite benefits would follow. This, for Corbett, was to confuse cause and effect. He did not deny that naval battle was important — he thought it was most desirable if it led to success. But what if it might not lead to success? Or what if the enemy stayed in port and declined battle? In the meantime, war went on and sea power had to be used to achieve military, trade and diplomatic objects. Sea power was control of sea communications (which at sea are parallel as opposed to being perpendicular on land) to enable a navy to do its will as far as the sea permitted it. For instance, so long as dispositions versus invasion were secure in 1805, it was still possible to pass a small army to the Mediterranean without a battle. This point could be enlarged. But the fact is that Corbett's view seemed to downgrade battle and the sailors distrusted it. To take the forces away from such fighting was to injure the 'spirit' of the service. It was somehow suspect. When in 1916, people were still unhappy about the lack of decisive success at Jutland (and before) Corbett was attacked for having induced the complacent mood that made such tame sea war possible. Actually, all that happened was that the Germans had acted as Corbett had predicted that they might. However, none of this alters the fact that his ideas of strategy on this main part did not capture the professional officers in the Royal Navy.

In the same way he failed to gain strong support for the idea of service co-operation. The navy men were bored, by and large, by the idea of transporting sea-sick soldiers about on wade-ashore missions. They were not easily able to understand the value of conjunct effects. To use a battleship as a troop convoy attendant was unpopular — although it became necessary when the Daughters of the Empire answered the Imperial call in 1914. In any event, no practical practice sessions or realistic plans derived from joint training experience occurred on the practical level. This was the negative result.

What about the gentlemen in higher latitudes? Corbett, as you all know, was a friend and advisor to Admiral Fisher. He supported that unorthodox sailor's matériel and administrative reforms in public. There is plenty of evidence to suggest that Fisher was no strategic ignoramus, and he knew some history despite the fact that he used to refer to history as "the record of exploded ideas." The problem with Fisher was that his matériel changes had aroused such opposition within the service that he could not risk making his strategic plans (if he had any) public where they could be used to shake confidence in himself and the Admiralty. There was no naval staff. In 1907, he had a naval war plan drawn up. It was derived from a war game played at the War Course and from Corbett's view of strategic principles. Probably it was never taken very seriously and was cynically concocted so that Fisher could, if questioned, say such plans did, in fact, exist. Meanwhile, when asked where the war plans were he would tap his head significantly and mutter "safe in here." As for Corbett's 'preface' to these red herrings they did not cause any revolutionary shock to penetrate the keep of the Admiralty. When Sir Arthur Wilson[12] was commenting on the war plans as a whole he minuted that any reader could skip Corbett's preface since that was only theory based on history. Quite a fall for a document of which Fisher had written "It is the epitome of the art of war! It's going to live!" When one considers that, when Sir Arthur Wilson was faced with explaining the Navy's strategic role to the CID in 1911, he could only stutter out salty inanities while General Sir Henry M. Wilson, for the Army, ran verbal rings about him, one can appreciate that perhaps a little strategic and historical reading would have done him some good.

It is true that Winston Churchill established a thing called a Naval Staff at the Admiralty before the First War. But even he was forced to concede after the war that his naval officers were more Captains of ships than Captains of war — recognizing that it takes years to train men's minds.[13] By 1935, Corbett's ideas were permeating the Naval Colleges and young men's minds: at least one war out of date. In 1912, when Captain Thomas Troubridge succeeded to the job of Director of War Plans Division, the appointment sounded grand. But Troubridge soon found out that he was shut out from the naval power structure. The cock ceased to strut. As Corbett shrewdly noted — "all the service is laughing at Troubridge." The navy was certainly not trained by 1913-14 to think strategically, let alone understand its historian tutor. I am not prepared to say whether that was for the best or not. But in the conflict between matériel mania and service politics, Corbett's thought was a casualty and, right, or wrong, it was all that existed to raise such considerations above the mundane. The voice of history was rejected.

No doubt there will be different views of whether or not it was or is wise to dispense with naval history. I am easy about it. But it will be well for you to remind yourselves that the voice of history was also the voice of intelligence. On one occasion I heard a great air commander contemptuously tell an audience that nobody paid attention to historians, that it was the ones who were there who counted. His audience contained such men as Arthur Marder, Noble Frankland, and Gordon Craig. It makes you think.

When the war came, Corbett was taken in to the Admiralty as a civilian to draft memoranda for the Cabinet. This was because he had made one convert — M.P.A. Hankey of the Royal Marines Artillery and then Secretary of the Committee of Imperial Defence. Indeed Hankey was convinced that it would be a war of amphibious operations, i.e., that Corbett's ideas had won. He was quite wrong. The guns in France won over everybody. *They* were not part of *any* strategical plan.

The Navy in the war played — after the first months and excepting Jutland and Gallipoli a fairly quiet rôle — a rôle based more on what Corbett said was likely than on what the sailors prepared for. The fleet performed creditably, and naturally enough ignorant politicians and journalists thought it did nothing while all the while it exercised the command of sea communications.

Corbett's strategy did not convince the British. That it was based on history, and that it was the only support to war games and common sense is indubitable. As Lord Esher once said — "Julian Corbett writes one of the best books in our language upon political and military strategy. All sorts of lessons . . . may be gleaned from it. No one, except perhaps Winston, who matters just now, has ever read it. . . . Obviously history is written for schoolmasters and arm-chair strategists. Statesmen and warriors pick their way through the dusk."[14]

It can be appreciated from the foregoing that somewhere along the way towards describing Corbett's strategic ideas I stopped. After all, his main strategic ideas he set forth in a book and in the pamphlet that I have mentioned. Also, like the pleasures in *Whisky Galore* by Compton MacKenzie, Corbett's strategic ideas are few and simple. What interested me more and more as I progressed with this paper was the relationship of military history to the popular war cry of the age — relevance. Speaking personally, and perhaps arrogantly, I have always avoided this question because it has never really much interested me as a practicing teacher and writer.

Apart from my own idiosyncracy, however, there is no reason why it should not interest others. The political strategist, military or civilian, has every right to ask — Of what value is history to the study of modern military diplomatic problems? I want to discuss this for a few moments with Corbett in mind.

First of all, I wish to emphasize that what I intend to say is rather subjective. Therefore, it is necessary to exempt all those of my historical brethren for whom quantitative analysis holds great and no doubt rewarding charms from my remarks. Verily, they have their justifications and their rewards.

Let us look at Corbett, whom I would not patronize by trying to justify. I have already pointed out that his first and most formative historical work was written with the idea in mind of attempting to find out the national significance of Francis Drake — based on a study of the records. I would be the last to deny, as I have indicated, that the study of Drake induced the idea of a general pattern into Corbett's mind and that he drew 'lessons' from it. Those lessons he passed on to others during his increasing intercourse with naval minds. Like so many of us, he found in history a certain knowledge that could be generalized for instructional purposes, or given precision for point-making when he was engaged in an argument with the then modern strategists. This represented a kind of self-educative process that allowed his mind to dominate those less well-informed minds with which his duties as War Course Instructor brought him into contact. Whether domination meant communication was, as I have indicated, another matter. Indeed, it is very difficult to know just how far the communication went. After years of instruction he potted his knowledge into *Notes on Strategy,* and again as I have indicated, Admiral Custance pointed out that the method was faulty — the right way was to induce his students to read for themselves.

This brings us to his books. Of what 'relevance' were they? It seems clear that as purveyors of immortal truths, truths that could be culled and presented for instructional purposes, they were of little value indeed. This was true for those times. It is even more true for today. The basics, themselves, could be extracted by a clever advocate of simplification, but how could they hold real meaning for those to whom the complicated total picture was only half guessed at or understood? And one must admit that Corbett, expert writer that he was, subjugated his materials to a developing purpose that was inevitably less than the whole picture. He did not aim at disinterested perfection, so much as he aimed at giving form and shape to what his superior mind perceived. In this process, lack of prejudice and impartial standards had only a limited rôle to play.

Then why read his books? The answer seems to be if one wishes to find the main ingredients of his thought in a hurry — don't. For the fact is that Julian Corbett was an artist. Like most artists he possessed skill — technical skill, in the use of his materials. One can admire the brushwork. But, more important, this artist produced an overall effect — one that emerges gradually as the beautiful symmetry of the grand design sinks into the consciousness of the reader. To partake of this benefit one needs both to savour the work in question with care, deliberation, sympathy — yes and perhaps even love — and then, as Wordsworth would say, to recollect in tranquility. This kind of suspended judgement and investment in time and reflection is necessary or the value of a Corbett book is lost. There is more to it than appreciating a full artistic effect. There are the side-effect connections that were only dimly made in the mind of Corbett himself and which only careful thought can either establish or begin to appreciate.

This is not intended as hagiography. In many respects Corbett was an obscure writer. Sometimes he appears to have been wilfully so. Sometimes he was careless with his consecutive thought processes — sloppy. Sometimes he was guilty of arranging facts to achieve an artistic effect that does violence to our sense of the credibility of his historical argument. As his son-in-law Brian Tunstall observed, he was a difficult writer — food for specialists. It is not difficult for me to see how a good mind — a mind practical and full of the desire to categorize, remember and obtain food for instruction — would reject such a book. He would recoil at investing time with the work of one who combined erudition, dreams and historical fact. He might recognize that the sentences are good, the material is intriguing, but, conclude finally, too obscure. He might also, and perhaps he might not admit this, find it all too complicated — too real.

Now, as I say, I have no wish to either 'defend' Corbett or to patronize him. What I have tried to do is use him as a vehicle for many, not all, historical writers. For it is my conviction that historians divide themselves into two groups — the relevant ones amenable to practical questions and the artists who use history suggestively. Of course there are shades and combinations of the two. For the man who seeks after relevance, it often appears that the 'practical' practitioners are the reliable truthful ones. But I am not so sure. Artistic performers like Corbett have the wit never to confuse the approximation of truth with truth itself and that kind of honesty is useful in an age of instant productions. There are, in fact, many men still writing, historians like Corbett, who fit this description. Some of them are here at this conference. I am talking about men who set out as research students and who get caught up not merely as professional historians, but as artists. The fact is that they are going to continue to exist and those who go to them for "their basic ideas" or "the main thrust of their argument" will come away nearly empty handed — grumbling about impractical or irrelevant stuff or about elegant antiquarianism.

Now that is as far as I want to go. Such men are never going to be relevant in the modern sense. They are not, and never can be "students of strategic studies." They mostly would not want to be so regarded, indeed they would be bad at it. As Leonard Beaton[15] once said to me, they often haven't done their hardware homework. Quite right. But all this is far from saying that the modern man, moved by practical necessities and who hears time's winged chariot hurrying near, would not derive benefit from reading and chewing on the artists or would-be artists I have been describing. It may be that I am wrong in my judgements that Corbett was a great artist. That is immaterial. There are others to choose from. Such persons stand four square — they condescend to no one, they really need no defence but they are *not* and never can be cogs in the wheels of defence machinery. Finally, if they are neglected by the busy men who are active in promoting a marriage between immortal truth and the latest thing in twenty minutes, then it may not be the historians who are the losers.

2 German Seapower

A Study in Failure

THEODORE ROPP
Duke University

Though this is the centenary of the Second Reich's first naval programme, this sketch of German sea power covers only half of that time. It centres around five men: the sea-minded Emperor William II; Alfred Tirpitz, his Naval State Secretary, 1897-1916; Erich Raeder, Naval Commander-in-Chief, 1928-1943; the land-minded Adolf Hitler; and Karl Doenitz, who surrendered what was left of the Third Reich in 1945 in the area where his Navy had been founded and off which it had fought its one big fleet action. It may also suggest (1) how a set of general ideas — such as those of Alfred Thayer Mahan, Guilio Douhet, or J.F.C. Fuller and B.H. Liddell Hart — may be (2) sold by a military interest group to (3) a particular state in particularly "ripe" circumstances, and (4) how that group's institutionalization of those ideas may (5) affect military operations. Though Tirpitz was not responsible for operational planning, which was under a separate High Command established — along with a separate personnel office — in 1889 in one of William's first acts, there were no rifts over planning. From his sale of the Fleet Laws of 1898 and 1900 to the Reichstag, acts which won Tirpitz his "von," he dominated the German Navy.

One asset which he had used and which was to protect his programmes later was the Navy's position as the only national armed force in an Empire whose oldest national flag was the North German Confederation's naval tricolor of Hohenzollern black and white and the red and white of the Hanseatic League. It reappeared in one quarter of the new naval ensign, with the Teutonic Order's Iron Cross in its centre and a black eagle in the centre of the black cross which divided the rest of the white field. The Imperial standard, designed by the Crown Prince and later Emperor and King Frederick III, was described in the 1910-1911 *Encyclopaedia Britannica* as one with "the iron cross with its white border displayed on a yellow field, diapered over in each of the four quarters with three black eagles and a crown, surrounded by a collar of the Order of the Black Eagle. In the four arms of the crown are the legend *Gott mit uns 1870,"*[1] as a reminder of that atmosphere of absolutism, militarism, and popular nationalism in which Germany's military programmes were sold.

But new arms programmes may demand resource commitments which run against traditional mind-sets, and time to carry them through. Tirpitz got both under a constitution which preserved the Emperor's control over internal and national security and foreign policy from more than budgetary obstruction by a Reichstag elected by universal manhood suffrage. The Chancellor was not responsible to the Reichstag; the Army's lack of a State Secretaryship happened to make it harder to cut back a naval programme which had won a long-term grant from that body. Long-term arms programmes, like the German naval one of 1872, were not new. They had long reflected rulers' efforts to ensure security by building fortifications — a common social investment for agrarian societies with seasonally unemployed labour — and to regularize their demands on their peoples by fixing their requirements for trained men, weapons, and material stockpiles. Modern peoples' wars had changed legal forms. The German Navy wanted a variant of the Army's Septennats of 1874 and later, which gave it one percent of the population and a fixed sum per man for seven years. But many naval programmes foundered in an era of technological uncertainty; in 1886 Lord Brassey's first *Naval Annual* gave Germany only one first class battleship, the 1868 *König Wilhelm*. In asking for a twelve year term in 1874, Helmuth von Moltke, the Chief of the Army's General Staff, had argued that such commitments were internationally stabilizing; any reduction would "bring uncertainty into all the many comprehensive preparations which must be made long in advance."[2] More than two decades later the coincidence of a new certainty about a battle fleet's ability to win and exploit the command of the sea and a naval race which might leave Germany as its colonial power broker gave her a chance to exploit the resulting international uncertainties.

The Progressive Eugen Richter's prediction that 1874's "bit of absolutism will eat its way forward like a cancer"[3] has been studied by many recent historians. But there have been few works on German navalism since Eckart Kehr's 1930 *Schlachtflottenbau und Parteipolitik.* Johnathan Steinberg's study of the 1898 Fleet Law criticizes Walther Hubatsch for blaming technology — that *diabolus ex machina* for many recent writers — and the claim that "The Navy Law like the Schlieffen Plan and the clockwork mechanism of mobilization became independent of the necessarily flexible diplomatic and political influences," but the German Navy has had no Arthur J. Marder.[4] While nobody

13

has calculated the size of Germany's military interest groups, one indicator for naval interests — as Mahan noted in his 1890 list of "the principal conditions affecting the sea power of nations" — is "the number following the sea, or . . . available . . . for the creation of naval material."[5] In 1912 Archibald Hurd and Henry Castle put Germany's "maritime population" at 80,000, shipyard personnel at 89,947, and the Navy's at 66,783, in a population of 65,400,000. Philip A. Ashworth noted in the *Britannica* that the Army's 1910 strength was 615,000, and that there had been 52,104 university students and auditors in 1906-1907. So British writers tended to agree that Germany's sea power was — by Mahan's standards — artificial. "Should the international horizon clear," Hurd and Castle concluded, "German politicians may, in spite of the insidious . . . Tirpitz, . . . awake to the [fact] that . . . their recent naval policy . . . [has] weakened . . . [Germany] by raising up opposition to her abroad and encumbering her with debt, taxes and Socialism at home."[6]

The problems of assessing the influence of those ideas which Mahan found in Theodor Mommsen and made into a general theory seem analagous to those of literary history. But such analogies break down when military ideas are institutionalized. And while civilian bureaucracies, such as those dealing with public health, are subject to "outside" shocks, such as those of the cholera epidemics of the early nineteenth century, the discovery — by statistical analysis — of its social before its natural causes, depressions, and wars, these establishments may be less tightly controlled than military ones, and their shocks less severe and sporadic. In any case, such major jobs as the creation of a battle fleet or an Air Force may increase the size of the bureaucracy which plans to secure "the specific powerlessness of a given enemy." In 1882 some 663 German naval officers were budgeted for active duty. By 1897 there were 761. The figure for 1914 included marines, but naval officers had more than quadrupled in a total of 3822. The 3229 Royal Naval officers budgeted in 1897 had increased by two-thirds to 5283 for March 31, 1914. The not quite comparable 1885 figure had been 2750. From 1897 when United States Naval officers on active duty had reached a post-Civil War low of 1399, they had more than doubled to 3406 by 1914. From hindsight into a United States Army Air Corps which expanded into an Air Force and from 1600 to 48,497 active duty officers from 1939 to 1948, the expansion of the German naval establishment promised optimal promotions, heavy workloads, and minimal reexamination of its agreed strategic concepts.[7]

The 1930s, when we began to mine these mountainous materials, are now as distant as we then were from the 1890s, but their glimpses into what Carolyn E. Playne called *The Neuroses of the Nations* are still as fascinating. The German Navy's "ideal," Hurd and Castle felt, was rather like "that of the American manufacturer. . . . He provides a motor-car . . . which will run well for a reasonable time, . . . at a price which justifies scrapping . . . [for] a gain in power." The

German navy would "look well on paper, ... and thus achieve a diplomatic objective," and would "be trained ... for a sudden coup" because of the disadvantages of a conscription system under which a fourth of its crews would be "raw recruits." Hurd and Castle knew that "the inherited sea habit counts for less ... in the handling of complicated mechanical appliances," than the value of the German "compulsory system of education," and that German officers worked as hard as they worked their men, but British "long-service men" would show a higher "moral standard" in wartime.[8] Marder has shown the solidity of the Royal Navy's faith in itself, while other historians have shown that of the Great War conscript armies of the Third Republic, Second Reich, and British Empire.

Another now noteworthy feature of these naval establishments was their small numbers. In 1898 the Royal Navy had fewer officers than the 1971 teaching staff of the University of Toronto (3329 to 4477); its 1914 staff was under Harvard's (5283 to 5700). The German Navy's 1898 staff was just below the 1971 faculty for the University of Western Ontario (761 to 1000). So a Fisher or a Wolseley crowd is very much worth studying, though Tirpitz was more important in history than any single British professional officer. Hurd and Castle saw him as "the nearest approach to a really great man that Germany has produced since Bismarck."[9] Of the fifteen men for whom National Socialism's heaviest ships were to be named at one time or another, only four — Albreckt von Stosch, Tirpitz, Reinhard Scheer, and Franz von Hipper — had been more than tangetially connected with the old Navy. We know that it had not overcome what Mahan had seen as unfavourable "natural conditions" and what he might have seen as the "unwise action of individual men," who have "at certain periods had a great modifying influence upon the growth of sea power."[10] A second-best German battle fleet had made some political and military sense in 1897, but had lost much of that within a decade, and was always opposed by the Army. This returns to the question of how navalism not only took root, but survived in such barren soil, and suggests some other questions. Are naval or air armadas more destabilizing than ground ones, since they are more generalized power projectors? Are they more easily sold by such generalized slogans as "a place in the sun?" Do their ties with more specialized industries give them more economic and political leverage than armies, which may also be more dependent on the blood tax of conscription? Or was German navalism only the most spectacular case of a mania which became epidemic when battleships were prestige symbols, one whose delusions were enhanced by political and military structures which limited Germany's ability to make "rational" military choices?

What happened is familiar. Before anyone saw that nobody was running a state whose "efficiency" — Mahan wrote in his 1911 *Naval Strategy* — "is so greatly superior to that of Great Britain, and may prove to be to that of the United States,"[11] that splitting, which Tirpitz was to find "almost fatal in war," had frustrated every effort to cut down his programmes. Germany's fleet, he had told the Reichstag in 1900, would not have to equal "that of the greatest naval Power, ... [which] will not, as a rule, ... [be able] to concentrate all its ... forces against it," or defeat it without imperiling its "own position in the world." The German fleet's function was political. As he had put it in a secret memorandum in 1897, "England ... [is] the enemy against which we most urgently require ... naval force as a political power factor."[12]

After William's 1890 "New Course" had unwittingly sponsored a formal Franco-Russian alliance, Britain faced the Dual and Triple Alliances and the growing United States and Japanese navies. By 1897 Germany might hope to tip the balance with a locally powerful fleet — the destabilizing factors in Gerald S. Graham's brilliant summary of *The Politics of Naval Supremacy* — with, as Tirpitz put it, "its greatest military potential between Heligoland and the Thames." He had already noted that "commerce raiding and transatlantic war against England is ... hopeless, because of the shortage of bases on our side and the superfluity on England's."[13] By 1914 Germany had the second largest navy in the world, but Britain's was also concentrated in the North Sea — as Mahan had noted in his *Naval Strategy* — to cover her "lines of communication ... practically with the entire world" and her "Islands against an invasion in force," and to cut "all German sea communications except with the Baltic." This had caused a minor revision of Tirpitz's "risk theory." When the British fleet blockaded Heligoland, German destroyers, submarines, and mines would weaken it before a fleet action in which Germany's shorter range ships would be tactically superior. The Index in Mahan's 1890 *Influence* had no headings for alliances, balance of power, or Germany. But his 1911 *Naval Strategy* "emphatically" revised his 1897 "opinion ... that European politics are scarcely ... part of the [United States Naval] War College course." Germany's "industrial, commercial, and naval" growth and military "pre-eminence," Russia's weakness, France's population gap, and close Austro-German ties meant that "the power to control Germany does not exist in Europe, except in the British navy." The "power of the German navy," especially in relation to Latin America, had become "a matter of prime importance to the United States."[14]

Every schoolboy now knows the names of such Wilhelmine scientists, engineers, and businessmen as Rudolf Diesel, Fritz Haber, Heinrich Hertz, Hugo Junkers, Max Planck, Walter Rathenau, and Ferdinand von Zeppelin. But in spite of their eminence in science, the Germans did not lead the British in its practical application to warfare. That specialization in research which had made Friedrich Bayer's dye fixant a universal pain reliever was often limited by the lack of overall political, military, scientific — Haber was a wartime army captain — staffs and planning bodies. Chancellor Theobald von Bettham-Hollweg later claimed that he had never looked into the army's plan to swing through neutral Belgium, and Alfred von Schlieffen's and Tirpitz's military dreams were never subjected to outside analysts. Tirpitz was a former destroyer man, but submarines were no better handled with his fleet than airplanes were handled later. "History," Mahan remarked in his *Naval Strategy,* "gives ... the qualifying factors; whereas reason, in love with its own refinements, is liable to overlook that which should modify them."[15] Tirpitz's navy never even saw the problem of forcing the British to come to Heligoland to be cut down to size. After taking over the state in 1916, the Army had to gamble on submarine commerce raiding, but what Tirpitz saw as an effort which had been aborted by political weakness ended with mutinies in the battle fleet, revolution, and surrender. His fleet's last victory was its scuttling off Scapa Flow.

Raeder was to hold his office almost as long as Tirpitz. The Navy had to live down the memories of its mutinies, its complicity in the failed Kapp *Putsch* of 1920, its taking in of the naval *Freikorps* units which had participated in it, and a captain's leaking the clandestine rearmament projects of both services. It welcomed Hitler more openly than the Army, but Raeder became a war criminal without ever becoming one of Hitler's paladins. Raeder's two volumes on cruiser warfare, written after the Kapp *Putsch* had pushed him into the naval history office, won him an honorary degree at Kiel, and may have been one source of his plan for long range raiders as the eventual core of a new battle fleet. But his selling this plan to the land-minded Hitler was aborted when the latter decided for war in 1939 after assuring his people that one would not break out before 1944 or 1945. Raeder's balanced force of eight battleships, five battle cruisers, four aircraft carriers, and 250 U-boats would have been far over the limits of the 1935 Anglo-German Naval Agreement, but "political negotiations were to [have been] launched at the appropriate moment."[16]

Though all planners had been so subjected to outside shocks that only the Maginot Line and two Russian Five-Year Plans had been completed by 1939, the German Navy had done little to secure cooperation with the localized fleets of Italy and Japan, or with the German Army and Air Force, while a bigness bug which Tirpitz had never caught and a destroyer shortage hampered efforts to help the Army in Flanders and the Baltic. Doenitz had argued that submarines would be easier to divert into other closed seas, but he had also found only what he wanted to find in history. Philip K. Lundeberg notes that he had not looked into more than the first two volumes of the "notably dispassionate" history of the U-boat war because he had found it "essentially negative." In spite of "the fretful shadow of Tirpitz," that history had "revealed the inadequacies of German planning and U-boat construction programmes, the fatal consequences of a disregard ... of wartime diplomacy, and ... [some of] the strategic consequences of chronic command disunity." But though it "provided invaluable guidance for a revived U-Waffe regarding unity of operational command and a broad doctrine for strategic deployment, ... this mighty disaster" was never "related to a discernible grand strategy."[17]

Doenitz was as small-minded as his predecessors. He blamed the *Bismarck's* "overdiversion" of British resources for her loss, but used the same line to justify otherwise unacceptable U-boat losses before the Allies' cross-Channel invasion. Three years earlier he had argued against diverting U-boats to Russian waters by claiming that, "The decisive factor . . . against Britain is the attack on her imports. . . . [This] is the U-boats' principal task and one which no other branch of the Armed Forces can take over from them. The war with Russia will be decided on land, and in it the U-boats can play only a minor role."[18] The Murmansk convoys were not large targets, but this "underdiversion" of German naval forces may possibly have affected the Battles of Leningrad and Moscow, as it had possibly affected that of the Marne, and certainly affected the Dunkirk operation. Tirpitz's High Seas Fleet had turned Britain's attention to closing her sea — one about as wide as Lake Superior is long — by better interception forces. His idea that "against England, indeed against any forces penetrating our home waters, the value of scouting vessels is much reduced," ignored the fact — which Mahan noted in discussing defensive operations — that an inferior force must find an enemy detachment to trap it.[19] And the British eventually used their scientific skills to find, fix, and strike submarines as effectively as the Germans used theirs to coordinate their attacks.

The main themes of this Anglo-German Thirty Years War are of equal skills in adapting to new sea, air, and ground fighting technologies, and of superior British political "efficiency." The failure to re-examine the strategical implications of the political decision to build the High Seas Fleet was as much institutional as individual, though military men may be especially prone to find failures of nerve rather than those of imagination, insight, or institutions. Some of this story may also be relevant to some of the problems of a decade in which a reduction and regrouping of United States forces may go with continuing discoveries of fuels and materials sources within their power orbit, substitutions for scarce materials and bases, and a growth of Russian sea power. Over a period which is now as long as that from 1887 to 1914 the Russians have built a conventional and a missile submarine force, and expendable mercantile, conventional naval, and ideological forces to close seas and deny resources in cold or conventional wars. The resulting sea power interests may be harder to freeze in disarmament talks than those related to ground or air power. Russo-American nuclear and air parity already exist. American ground and Russian naval inferiority are based on positional limitations. But Russian ideological and economic growth interests could rally around naval forces which could enhance her political position in various areas, especially in an Indian Ocean which, with the loss of Western military transit rights through the Arab world, is again as close to Bermuda as to London. So the race for sea power could continue, while those for air and ground power slacken.[20]

3 The R.N.A.S. in Combined Operations 1914-1915

COMMANDER W. A. B. DOUGLAS
Directorate of History
Canadian Forces Headquarters

War tends to shatter preconceived notions, and those concerning the Royal Naval Air Service in 1914 were no exception. The principal function of naval aircraft was to scout for the fleet, but the rapid unfolding of events thrust naval aviators into a number of unexpected situations during the first fifteen months of the war. Consequently, tactical and technical developments took place that put the R.N.A.S., in nearly all respects, far ahead of any other naval air service. The exception was in rigid airships, where Germany always led the field. Then, in 1916, R.N.A.S. development slowed down to such an extent, that by the end of the war British naval aviation still had not advanced very far beyond the experimental stage. Now it is possible to advance very good technical reasons to show why aircraft capabilities in support of the main fleet were strictly limited. And like the first official airforce historian, Sir Walter Raleigh, one may indulge in veiled[1] or open[2] accusations of prejudice, within the Admiralty, against the R.N.A.S. Such explanations do not do justice to the exploits of the R.N.A.S. between 1914 and 1916.[3] Nor do they really help us to understand why the Admiralty became so committed to long-distance bombing, although that is another story. It is the purpose of this paper, then, to focus on the activities of units under the command of C.R. Samson and, later, F.H. Sykes, in Flanders and the Eastern Mediterranean between September, 1914 and January, 1916. In these operations sea, land and air forces acted in conjunction against the enemy. Although ships took little part in the Flanders campaign, the developing role of the naval air service offers us the first known example of air forces working in cooperation with other services. How the R.N.A.S. became involved, the nature and effectiveness of its operations, and the influence of the experience on naval aviation, are questions that have been overshadowed by a universal preoccupation with the fascinations of Grand Strategy.

I

Under the protection of Winston Churchill, the Royal Navy had acquired the largest naval air service in the world by mid-summer of 1914. Aeroplanes and seaplanes predominated, although most forward-looking professional seagoing officers wanted airships.[4] Many of the aeroplanes were prototypes for seaplanes, but Churchill fostered them to defend important targets against German air attack.[5] Before the war he encouraged Wing-Commander C.R. Samson to experiment with the Eastchurch aeroplanes in bomb-dropping and aerial fighting,[6] gave the unit a roving commission, and sent Samson to Ostend on 27 August, supposedly to provide aerial reconnaissance for the Marine Brigade.[7]

Nine aeroplanes, most of which had flown at the naval review in July, "old veteran servants of the crown," flew to Belgium. H.M.S. *Empress,* not yet converted to a seaplane carrier, sailed in company with a collier from Sheerness carrying ground personnel and a remarkable collection of motor cars and buses.[8] The most up-to-date of Samson's aeroplanes was a Sopwith three-seater tractor biplane powered with an 80 H.P. Gnome engine.[9] He also had four other aircraft powered with the 80 H.P. Gnome – a Bleriot monoplane, a Henri Farman biplane pusher, a Bristol biplane and a Short biplane (converted from a seaplane). One Bleriot monoplane with a 50 H.P. Gnome and two B.E. biplanes with 70 H.P. Renault engines – including Samson's favourite, number 50 – completed the "aeroplane party," as the Admiralty called it.[10]

It is hard to believe that the Air Department intended this extraordinary establishment simply to provide the Ostend diversion with aerial reconnaissance, particularly since the Eastchurch squadron was the backbone of England's air defence, Lord Kitchener not having sufficient aeroplanes to meet the War Office's obligations in this regard.[11] It is important to realize that the Admiralty's approach to air defence, as demonstrated by a study of April, 1914, was to adopt a resolutely offensive strategy.[12] Thus it is not surprising that when the Admiralty did recall the Eastchurch aeroplanes, along with the Marine Brigade, on 30 August, Samson turned a blind eye and

> decided that there was going to be a bad fog in the Channel which would force us to land at Dunkirk. A suitable landing ground just west of the town near the suburb of St. Pol had seen selected and there we duly landed. The general commanding the garrison, having

been persuaded by Samson that we could be of some use to him, gave us the use of the ground and the buildings of a disused hospital beside it; so we settled in.[13]

Orders for Samson to come home gave way, on 1 September, to Churchill's instructions for the Eastchurch aeroplanes to search for and attack enemy airship bases. The intention was to form a more or less self-sufficient force, consisting eventually of three squadrons of 12 machines each, and one squadron of sixty "special motor cars," armed with maxim guns. Each aeroplane squadron was to work with an armoured car section, which would protect landing fields from enemy cavalry raids.[14] The "Dunkirk Circus," as it came to be known,[15] was buttressed by 450 men of the Oxfordshire Yeomanry and a brigade of Royal Marines from the Naval Division. After these troops had landed at Dunkirk, Churchill, on 22 September, authorized Samson to attack German lines of communication – an additional task that the armed motor cars had already been undertaking with élan.[16]

Circumstances never did permit the "Dunkirk Circus" to expand to its full establishment. Although its principal aim was to attack German airship bases – an aim that was achieved in spectacular fashion on 8 October by Flight Lieutenant R.L. Marix in a successful raid on the Düsseldorf sheds[17] – there is little doubt that Samson's chief intention was to get involved in the thick of battle with his armed motor cars. Responsible directly to Captain Murray Sueter, the Director of the Air Department, (D.A.D.), drawing supplies from the Central Air Office at Sheerness, and using H.M.S. *Empress* for sea transport, Samson created a rudimentary aircraft park with servicing facilities at St. Pol. Wing Headquarters, a mobile concept, invariably occupied the place most suitable for motor car operations.[18] Sueter reinforced Samson by the unorthodox method of telegraphing to Squadron Commander Arthur Longmore, at Calshot: "You are to raise all the aeroplanes you can and fly to Dunkirk by Sunday [27 September]. I give you a free hand to go to Hendon and Eastchurch and take any you like. Machines can either fly or go by steamer...."[19] Although he only received this telegram on Friday, Longmore was at Dunkirk by 3 P.M. on Sunday with six ssorted aeroplanes.[20] Eleven days later, with the only three aeroplanes he still had serviceable, he flew to Ostend to join Squadron Commander Richard Bell-Davies, who had brought five aeroplanes from Antwerp.

On 10 October, Lieutenant General Sir H.L. Rawlinson landed at Ostend with the 7th Division and Sir Julian Byng's 3rd Cavalry Division. Accompanying them was 6 Squadron, R.F.C. When it was clear that nothing could be done to save Antwerp, the two divisions formed IV Corps and fell back to take position on the left wing of the B.E.F., in front of Poperinghe. The Admiralty at this point recalled Longmore and Squadron Commander E.L. Gerrard to set up 1 and 2 (Naval) Squadrons in the relatively safe surroundings of Gosport and Eastchurch,[21] but Samson remained with his armoured cars, keeping Davies in command of what was euphemistically known as 3 Squadron. After a Bleriot monoplane crashed on take-off at Ostend, the squadron was left with two B.E.2a's and a Sopwith 80 H.P. Gnome.[22] In Samson's words, "Sir Henry told me in 1915, when I visited him, that he would never forget the way my aeroplanes had stuck to the IV Corps and carried out reconnaissances in filthy weather under the most difficult circumstances."[23] Others recalled the R.N.A.S. support of IV Corps in less flattering terms,[24] but the R.F.C. squadron lacked experience, and there is no doubt that Samson was in his element in this irregular type of warfare. His airmen learned valuable lessons at the Battle of Ypres, and his armoured cars were useful in the retreat. He later claimed that Sir John French ". . . wanted my party to join up with him, and have nothing to do with the military wing,"[25] a suggestion that would have further complicated his already curious command relationships. Churchill and Sueter gave him no further encouragement, and firmly, even though reluctantly, steered him back to the principal aim of his force.[26] Enemy vessels, submarine bases, and other targets on the Belgian coast were now to be attacked in addition to airship bases.[27] On 7 November, while Samson was still in England arguing his case, the last remnants of the "Dunkirk Circus" returned to Dunkirk. Thus, as the first battle of Ypres drew to a close, and static warfare began, the naval air service's first tentative foray into cooperation with troops and armoured cars petered out.[28] But neither Samson nor Churchill gave up hope of renewing the experiment.

II

The "expedition" launched on 1 September now became the Naval Aeroplane and Armoured Motor Car Patrol, and Seaplane Base at Dunkirk. The patrol was "under the general command of the Commander in Chief in France, who will authorize them to proceed with their special mission."[29] One of the three seaplanes[30] revealed a special virtue – it could carry a 100 lb. bomb.[31] The ideal aeroplane, naval aviators reckoned, should carry at least 300 lbs. of bombs at 90 m.p.h. or more. Nothing less than a 100 lb. bomb had any effect against submarines; and the only hope of success against land targets was steady and persistent bombing.[32] In February, 1915, co-ordinated aeroplane and seaplane raids against targets on the Belgian coast[33] showed the worth of 100 lb. bombs, the unsuitability of seaplanes for raids over defended targets, and the success of offensive tactics in establishing air superiority.[34]

By R.F.C. standards, these were rather minor operations. Twenty-one aeroplanes, seven seaplanes and eight French aeroplanes had taken part in the raids. On the Western Front, the R.F.C. had expanded into three wings and a headquarters squadron. The tasks of R.F.C. aeroplanes ranged from scouting to air photography, artillery cooperation and bombing railways in cooperation with the French, the latter role being strictly secondary.[35] On 1 January there had been six squadrons with 47 machines and by 31 March there were seven squadrons with 83 machines. It required roughly 12 new machines every three months to keep a squadron of 12 R.F.C. aeroplanes in the field.[36] Thus, even taking into account R.N.A.S. seaplanes being used for coastal and fleet operations, the R.F.C. had much the larger air effort. In order to satisfy R.F.C. requirements, the Admiralty had diverted aeroplanes intended for the R.N.A.S., and as a result of aircraft shortages Sueter reduced the coastal patrol of England to insignificant proportions.[37] Even so, aeroplanes, (and for the moment these were largely of French origin), held the field in R.N.A.S. planning. When Churchill cancelled the rigid airship program on 18 January, all the resources thereby diverted went to "the more practical aeroplane, in which we have been so successful." Later on, in April, he announced a target of 1000 aeroplanes and 300 seaplanes, and 400 more pilots, by the end of the year. His objective was a powerful offensive weapon, including torpedo-carrying seaplanes "for a night attack on German ships-of-war at anchor."[38] It was principally the units at Dunkirk, in spite of their motley composition, that had convinced the First Lord of the offensive possibilities inherent in the air weapon.

III

During this period, from January to April, the R.N.A.S. was committed to a new and demanding task at the Dardanelles. Dunkirk and Home Waters were still the centre of R.N.A.S. activity — indeed, the stimulus for Churchill's latest ideas came from Longmore after he had assumed the Dunkirk command.[39] Nevertheless, the Dardanelles exerted a fundamental influence on air policy. On 31 January the Admiralty sent *Ark Royal* with six seaplanes and two aeroplanes to join the naval squadron under the command of Vice Admiral S.H. Carden. On 26 February Sueter ordered 1 Squadron under Longmore to relieve Samson, and ordered 3 Squadron to the Dardanelles. *Ark Royal* arrived on station and began aerial reconnaissance on 17 February. 3 Squadron arrived on 24 March and began operations on 29 March.[40] It is clear not only from this sequence of events, but also from the planning arrangements, that the enhanced capabilities which air observation was expected to give to naval guns played an important part in the decision to go ahead with the naval bombardment. As Maurice Hankey, secretary of the War Council, told the Dardanelles Commission, ". . . in a comparatively confined space like the Gallipoli Peninsula, the value of naval bombardment, particularly by indirect laying, would be enormously increased."[41] The official historians, H.A. Jones and C.F. Aspinall-Oglander, have implied that this was behind the decision to send Samson's aeroplanes as well as the *Ark Royal.*[42]

Now it is true that Carden's replacement, Vice Admiral Sir John de Robeck, sent an urgent request for air reinforcements, but he did not do so until 9 March. By that date, as we have seen, events had overtaken him. The decision to send Samson had already been made on the same day that Churchill telegraphed to Carden that he was sending the Naval Division (including armoured cars) which he had recently discovered was not wanted in France.[43] Kitchener had refused the 29th Division for a possible combined operation at the 19 February War Council meeting, and he did not relent until the second week in March.[44] Even then, he refused flatly to spare any R.F.C. squadrons,[45] with the result that the naval air service provided all air support — just as it would have done in the eventuality of a combined operation that did not include the 29th Division.

There was no point in sending a boy to do a man's job, and Samson, with his experience in Flanders, his instinct for war, and the best mechanics and pilots in the R.N.A.S., was the obvious choice. He might not have been such an obvious choice if combined land and sea operations had not figured so prominently in Churchill's plans. Samson brought with him two B.E.2a's (including his famous number 50), two Maurice Farmans with 100 H.P. Renault engines, two B.E.2c's with 70 H.P. Renault engines, an old Maurice Farman 140 H.P. Canton Unné pusher — a veteran of 120 hours with the squadron in Flanders — a Breguet 200 H.P. pusher with an armoured nacelle, two Sopwith Tabloids and eight Henri Farmans with 80 H.P. Gnome engines, He had eleven pilots, two observers, one engineer officer, two doctors and 100 men.[46] On paper this was a useful contribution even by R.F.C. standards on the Western Front. Samson himself believed it was shamefully inadequate. He told Sir Ian Hamilton that he needed 30 good two-seater machines, 24 fighting machines, 40 pilots and 400 men. "So equipped, he reckons he could take the Peninsula by himself and save us all a vast lot of trouble."[47] In fact, of his 18 machines only three Maurice Farmans and two B.E.2a's were effective.[48] Tenedos, it must be remembered, was 17½ miles from Cape Helles and 31 miles from Anzac Beach. Every flight over enemy territory first demanded the equivalent of flying the English Channel.[49] Rabbit Island, halfway between Tenedos and Helles, provided no more than an emergency landing place covered with loose stones. A landing strip created at Helles after the landings was made untenable by enemy shellfire.[50] Dust flew freely in the everlasting wind, complicating problems of engine maintenance. The Henry Farmans were underpowered and difficult to keep in tune. The B.E.2c's were useless with a passenger embarked. The Sopwith Tabloids "had a habit of shaking out their engines," and the Breguet suffered repeated engine failures.[51] At the Dardanelles it was more the elements than the enemy that limited the effectiveness of aircraft.

By the time 3 Squadron arrived preparations for the Cape Helles landing were underway. With his incredible knack for improvisation, Samson created a photographic unit out of one private camera and its owner, and gave Hamilton the best intelligence available of conditions on the Gallipoli Peninsula.[52] He conducted a number of spotting flights for bombarding ships, and went directly to the offensive by dropping bombs on targets of opportunity.[53] *Ark Royal's* seaplanes had spotted mines and enemy entrenchments with great success even though they were unable to rise above effective rifle range.[54] On Samson's arrival Commander Robert Clark-Hall passed on the intelligence laboriously collected by seaplane observers, then, ". . . reaping the advantages of sea-power, went wandering off . . . to distant coasts, to the Gulf of Andromyti, to Smyrna, Enos and to Bulair. . . ."[55]

It was a soldier who now added a further dimension to R.N.A.S. operations. "A man-lifting kite or a captive balloon would be of great use to the navy," General Birdwood cabled to Kitchener on 14 March. "It would not only give great assistance in the spotting of long-range fire but I would also be able to detect by its means the concealed batteries which are already troubling the navy. . . ."[56] The navy had experimented with spherical balloons somewhat unsuccessfully. One of them, which had seen service in South Africa, was fitted in the tug *Rescue* in the Dardanelles and confirmed the unreliability of the type.[57] Late in 1914 French Drachen-type balloons proved effective on the Western Front, and in February the Admiralty had obtained two from the French government. Their Lordships conscripted the vessel S.S. *Manica,* which was unloading manure in the Manchester ship canal, fitted the necessary winches and gas plants, loaded aboard the two kite balloons, 6 officers and 83 men under the command of Flight Commander J.D. Mackworth, and sailed her to the Dardanelles on 27 March.[58] On 16 April Mackworth cabled home, "I consider that the results obtainable from the very rough and ready apparatus in the *Manica* warrants the question of properly designed kite balloon ships being taken up. . . ."[59]

By the time the first landings took place on 25 April the R.N.A.S. had thus made several significant contributions to the campaign, even though the air effort was disjointed. Clark-Hall and his seaplanes, Mackworth and his kite balloons, Samson and his aeroplanes, worked in almost watertight compartments. Vice Admiral de Robeck was the operational commander for all naval units in the theatre, but Samson had to work directly with the G.O.C., Sir Ian Hamilton, as well as the naval C.-in-C.[60] Furthermore, all R.N.A.S. units were still directly responsible to the D.A.D. for operations, and it was not until 1 August that the unequivocal statement appeared declaring the R.N.A.S. to be "in all respects . . . an integral part of the Royal Navy." Even then the air units in the Mediterranean were not mentioned, although it was clear that the Air

Department continued to require direct operational links with overseas R.N.A.S. forces.[61]

At the beginning, Churchill was the chief inspiration for the Dardanelles airmen, but in Hamilton's view "... they can't get the contact and they are thoroughly imbued with the idea that the Sea Lords are at the best half-hearted; at the worst, actively antagonistic to us and the whole of our enterprise."[62] Because Hamilton refused to go behind De Robeck's or Kitchener's back to Churchill, all he could do was attempt to support Samson's request to the naval C.-in-C. for more and better aeroplanes so that the R.N.A.S. could play a useful role in the expedition.[63] He was not altogether unsuccessful, and the Admiralty sent what aircraft could be spared. In May six new Henri Farman 80 H.P. Gnomes arrived (and were promptly returned, being more trouble than they were worth). In June five Voisin two-seaters with 140 H.P. Canton Unné engines arrived. Although slow and considered obsolete in France, they were good reconnaissance machines. Their lives were short because they had to fly continuously at full revolutions in order to maintain satisfactory height. It was in July that the first really effective reinforcements arrived, in the shape of six Nieuport Scouts fitted with 80 H.P. Gnomes. Samson glowed over the Nieuport – "it climbed like a witch." These aeroplanes were successful in single-seater reconnaissance, light bombing and aerial combat. Four Maurice Farmans arriving at this time gave Samson important additional strength.[64] The Admiralty also pressed on with kite balloon developments, following Mackworth's encouraging reports. A second ship, H.M.S. *Hector*, arrived in July and a third, H.M.S. *Canning* in October after *Manica* had returned to England for refit.[65] New seaplanes arrived for *Ark Royal* in April, more in May, and new engines (225 H.P. Sunbeams) in August. On 12 June H.M.S. *Ben-my-Chree*, a much faster seaplane carrier, arrived with the most up-to-date seaplanes available.[66]

Enemy air opposition and anti-aircraft defences did not pose a serious threat in 1915 to the R.N.A.S. or to the French aeroplane squadron that arrived at Tenedos on 15 May.[67] Significant R.N.A.S. developments thus tended to result from indirect or internal pressure. In France the success of Longmore, taking up where Samson left off, had a profound influence on future plans for Dunkirk and reduced the material available for Samson. In the North Sea, failures in the attempt to launch an aerial offensive from seaplane carriers resulted in the dispersion of such carriers. It was this that lay behind the decision to send *Ben-my-Chree* to the Mediterranean, to practice torpedo-launching against the Turk.[68] But these factors were no more than pinpricks in comparison with the effect felt from Churchill's and Fisher's resignations in May. Within days after the First Lord's departure the new Board had asked Jellicoe – a long-standing critic of Churchill's handling of the R.N.A.S. – for guidance on air policy. At the same time Fisher wrote to Jellicoe, "Had I been kept on, I had laid all my plans for getting back from the Dardanelles all our aircraft and reorganizing the East Coast air service so as to deal with the enemy's aircraft."[69] Jellicoe advised the Admiralty to draw in their horns. The functions of the naval air service, he pointed out, were observation duties from the coast, the attack of enemy aircraft, "the aerial defence of all naval centres, since the Army, who properly speaking, should carry out this duty, have apparently turned it over to the navy," and scouting for enemy submarines and minelayers. Or as Maurice Bonham-Carter said to Hankey, who agreed, the R.N.A.S., not having been designed for naval objects, "has degenerated into a crowd of highly skilled but ill-disciplined privateersmen." Disapproval settled about the naval air service like a shroud.[70]

It could have been as a result of this attitude that Colonel F.H. Sykes of the R.F.C., whom Samson detested, was sent to report on air operations at the Dardanelles.[71] Sykes recommended centralization, and other drastic improvements, that resulted in his own appointment in command; creation of a wing organization, making 3 Squadron a wing; the dispatch of a second Wing, (number 2 Wing, under Gerrard); the shift of base from Tenedos to the much more suitably placed Imbros; and the introduction of airships based at Mudros for anti-submarine patrols.[72] Centralization at the Dardanelles brought the airmen unprecedented independence. It is true that Samson led the way, but the independence he enjoyed did not extend beyond his own squadron. Central command at the Air Department afforded theoretical independence, but the D.A.D. could not advise or exert control like a commander in the field. It is true that similar developments at Dunkirk resulted in the appointment of Wing Commander C.L. Lambe to command all air units at Dover and Dunkirk, but he remained on the staff of a naval commander. Sykes held a position of great tactical advantage by providing the link to both a naval and a military commander requiring the services of a single air force. In addition, when the Admiralty decided to launch long-distance bombing raids against strategic targets not directly related to Dardanelles operations, the orders could be handled entirely by Sykes' own staff.[73]

The effectiveness of air operations at the Dardanelles has never attracted much attention. There were admitted limitations. Naval pilots and observers were not properly trained for spotting and reconnaissance,[74] and even when they became adept their usefulness was lessened by a wall of prejudice. Gunners in ships and batteries tended not to believe large corrections to the fall of shot; the air report that the Sulva landings were virtually unopposed resulted in no action by Lieutenant General Stopford.[75] On the other hand, aircraft of all kinds, (seaplanes, aeroplanes and kite balloons), directed some highly successful shoots, particularly against personnel.[76] On several occasions air reconnaissance reported Turkish counter-attacks with the result that the enemy troops were practically wiped out by artillery fire. The arrival of *Ben-my-Chree* resulted in having a seaplane carrier fast enough to steam in submarine-infested waters. Her torpedo-launching experiments were rewarded by a well-advertised success and several lesser-known failures.[77] Such trials were essential to progress in the field, and they were the only trials being conducted in the face of the enemy. Aerial bombing of Turkish communications (which included the first dropping of a 500 lb. bomb), had an effect second only to submarine attacks.[78]

Whatever the immediate results, the campaign permitted the R.N.A.S. in the Eastern Mediterranean to develop as a virtually independent air arm. With the arrival of enemy submarines, and the introduction of non-rigid airships, the forces under the command of Wing-Captain Sykes were covering the entire spectrum of military air operations. Sykes was aware of the significance of the situation, and sought to make the best of it.[79] Failure in the campaign, however, resulted in the Admiralty shifting support to Dunkirk and reducing strength in the Eastern Mediterranean. Even had this not happened, there is little evidence that Hamilton himself thought of the air weapon as introducing a new dimension to warfare. "Flying is not my stunt," he admitted, and his final dispatch, though eloquent, was hopelessly old-fashioned:

> ... These bold flyers are laconic, and their feats will mostly pass unrecorded. Yet let me here thank them, ... for the nonchalance with which they appear to affront danger and death, when and where they can. So doing, they quicken the hearts of their friends on land and sea — an asset of greater military value even than their bombs or aerial reconnaissance, admirable in all respects as they were.[80]

IV

It can be aruged that the experiences of the embryonic naval air service in 1914 and 1915 provided an important stimulus for future success.[81] In 1916 a force of 42 aeroplanes, nine seaplanes and three airships remained in the Aegean.[82] The Dardanelles adventure certainly provided the stimulus for that development. Some recognition may also have been given in the Admiralty to the appreciation by Captain R. Glyn of Sykes' staff, that "from a military point of view the future can hold nothing but extra responsibility for the air service. ... The Dardanelles operations have brought about such a close and happy relationship of cooperation between the two services, that it is now hard to classify any duty that the air service is capable of carrying out as belonging to either one or the other...."[83] On the other hand, the Admiralty failed to develop kite balloons, which had made such impressive advances, even though as Lord Sydenham observed eight months later, "the balloon was the only possible substitute for observation purposes for the rigid airships which the Admiralty denied to the navy."[84] Nor was anything done about torpedo-carrying seaplanes.[85] Rear Admiral Vaughan-Lee, who had superseded Sueter and become Director of Air Services in the new regime, did agree to station eleven Short 225 H.P. Sunbeam seaplanes at various stations on the east coast of England to launch torpedoes against enemy warships at night or while engaged by the fleet.[86] On 25 April, 1916, during the German naval raid on Yarmouth, the available torpedo-carrying seaplanes were not allowed to attack the enemy. A disgusted naval officer proposed that "pilots who have what is vulgarly known as guts [should] aim for results that might be of the very utmost importance. They would undoubtedly realize that, in company with a large proportion of the British army, they would be incurring a certain risk."[87] It was not to be. The first such attempt did not take place until September, 1917, under the direction of Rear Admiral Murray Sueter at Otranto, where he had been sidelined.[88] As submarines and rigid airships had already proven, weapon development was impossible without taking risks.

The immediate effect of the Dardanelles operations was, in fact, negative, because it marked the end of combined operations until 1918. Not only did the Admiralty cease to take a lively interest in kite balloons and torpedo-carrying seaplanes, but their Lordships disbanded 3 Wing, pushed Samson off to a sideshow in the Mediterranean,[89] and put Sykes at the disposal of the War Office. On the other hand, a new 3 (Naval) Wing was taking shape that reflected where the Admiralty's priorities now lay. At Detling, in Kent, a small group of pilots was training under Squadron Commander R.L. Marix, one of Samson's protegés and hero of the Düsseldorf raid. These pilots — nearly all recently qualified Canadians — would launch the naval air service on one of the first strategic bombing campaigns in history, from an airfield behind the French sector of the Western Front.[90] What combined operations both in Flanders and the Eastern Mediterranean had done was to accelerate development in several directions, persuading the Admiralty that the most fruitful results were to be obtained from the aeroplane, particularly in strategic bombing. Samson, through his innovative skill, and Sykes, through his clearsighted analysis of operations, were largely responsible for that achievement. When no further opportunity for combined operations arose, and both Samson and Sykes disappeared into limbo, the aeroplane received priority over other aircraft, and the air support of the fleet languished. It would take the stimulus of submarine warfare, and Beatty's search for an offensive weapon in 1917, to force the Admiralty out of its complacent neglect of naval aviation and restore the pace of technological development.

4 The Dardanelles Revisited

Further Thoughts on the Naval Prelude[1]

ARTHUR MARDER
University of California, Irvine

This paper is not concerned with the genesis of the Dardanelles campaign (though I consider it was the one imaginative strategic idea of the war on the Allied side) or the preliminary moves. Its scope is the naval facets of the opening phase of the operation, from the initial bombardment of 19 February to 25 April, when the Army took over the principal role, with the thrust on what went wrong and why and some second-guessing on what might have been done. The naval side of the Dardanelles is second only to Jutland in the longevity and passion of the controversy which it has aroused among naval historians.

The Turkish defenses at the Dardanelles consisted of four principal elements: the forts, the minefields (and minefield and mobile howitzer batteries), torpedoes, and floating mines. The last two were of secondary importance, though the moored mines exercised an influence on Admiral De Robeck's decision after 18 March.

It is the forts and the minefields on which we must concentrate. The intention at first was to overcome the forts with naval gunfire, *then* to sweep up the mines, and thereby open the way for the fleet to reach Constantinople. A short-range naval bombardment on 25 February silenced the forts at each side of the entrance to the Straits. Then came the bombardment of the intermediate defenses in the first days of March: the 36 mobile howitzers (mainly 5.9-inch) and 24 mortars (mostly 8.2-inch). Incapable of hitting a moving target, their function was to hit any anchored ships they could reach, so keeping them on the move. They were not intended to protect the minefields, which task was assigned to the minefield batteries. Though the bombardment of the intermediate defenses was not decisive, on 5 March the fleet initiated the next phase, whose objective was the silencing of the forts at the Narrows with their fixed batteries of heavy guns.

The Narrows forts could have been silenced by ships from inside the Straits by direct fire, but without ranging correction, unless they had air spotting. The alternative was to attack firing over the Gallipoli peninsula with a battleship inside the Straits to spot for range and an aircraft to spot for direction. (Using a battleship inside the Straits to spot for range was a refinement, but not necessary, as an aircraft could give both range and direction corrections.) Firing over the peninsula was the more promising way, since the weakness of the gun defenses at the Narrows lay in their vulnerability to indirect fire over the peninsula, though only the flagship, the *Queen Elizabeth*, had sufficient range for this. Firing from the direction of Gaba Tepe, she could bombard the forts with accuracy, *given efficient airplane spotting and sufficient ammunition*, since she could anchor outside the range of the howitzers. The guns of the forts, designed on the expectation of attack from inside the Straits, would present almost a broadside target to indirect fire over the peninsula. This was "about four times as favourable as the 'end on' target presented when attacking from inside the Straits. The forts were further unprotected against an attack of this description from the rear."[2] The European forts were particularly vulnerable.

With indirect firing, as in the case of a ship off Gaba Tepe firing over the peninsula, air spotting for direction was essential. The gunlayer could not see the target, and had to lay his gun on some object believed to be in the right direction, and then shift by sight-setting to right, or to left, as necessary, when he received corrections from the spotting aircraft. This elementary consideration must have been well known to any competent gunnery specialist, yet no tests or practices of battleships firing with aircraft spotting were carried out during the 17 days between the arrival of the *Ark Royal* at Tenedos on 17 February with six seaplanes and the time of the *Queen Elizabeth* firing over the peninsula on 5 March. The detachment of a battleship or two and a few gunnery officers to carry out essential spotting tests on a shore target, for which an unoccupied Aegean island could have been used, would not have impaired fleet efficiency or operations in any way. Such exercises were all the more necessary because the seaplanes were so few, and could be used only in favourable weather (they were generally unable to rise in any but smooth water). The Aegean weather in March was a definite limitation. "The heavy and cumbersome floats over-taxed the low-powered engines, and we were constantly pre-occupied with keeping our machines in the air," writes Williamson. Also, the wireless gear was somewhat unreliable. But such disadvantages were largely counter-balanced by the enemy having no aircraft and no anti-aircraft guns, while usually the seaplanes were able to fly at sufficient height to be untroubled by rifle fire. In

THE SEA OF MARMORA
AND APPROACHES

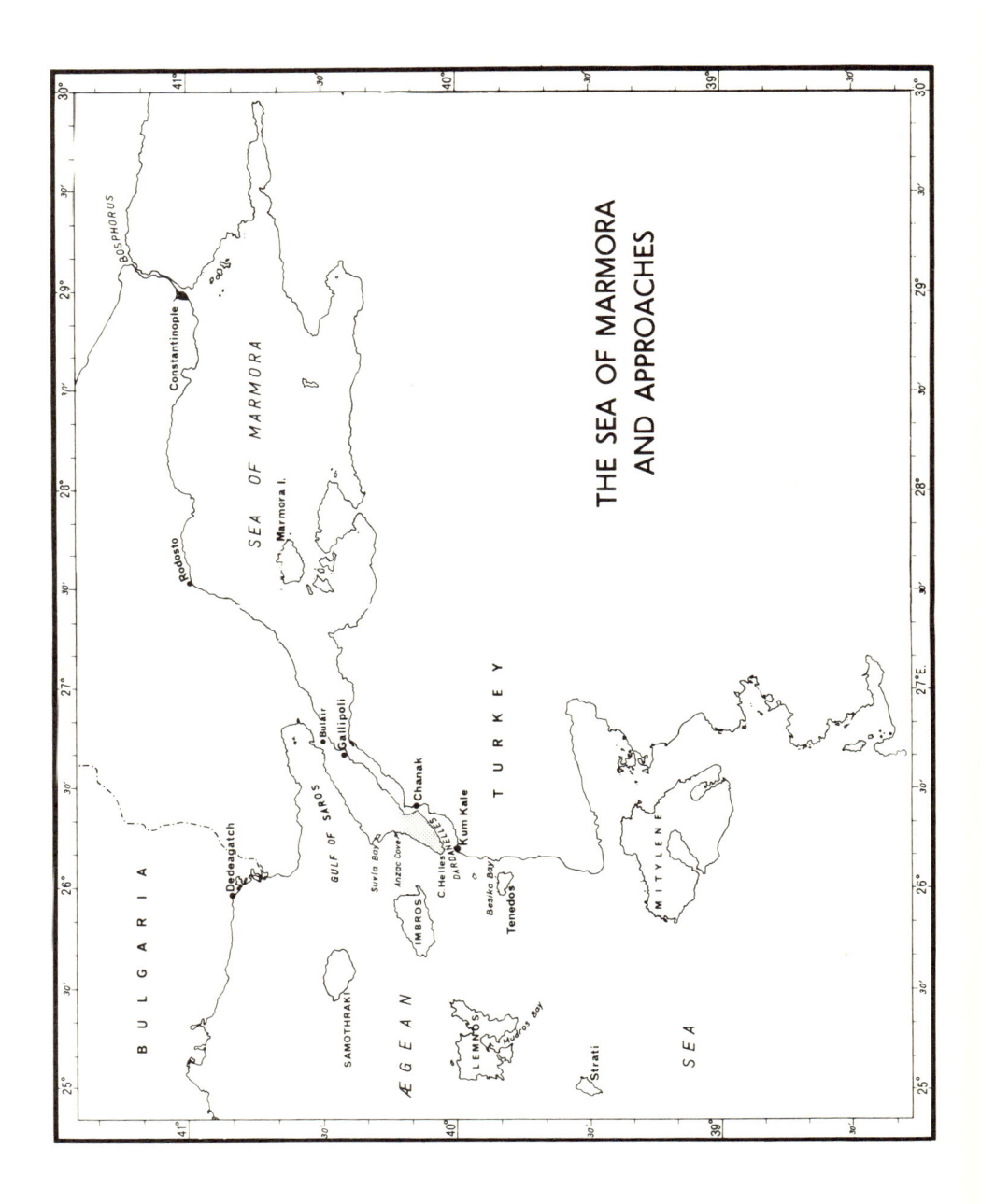

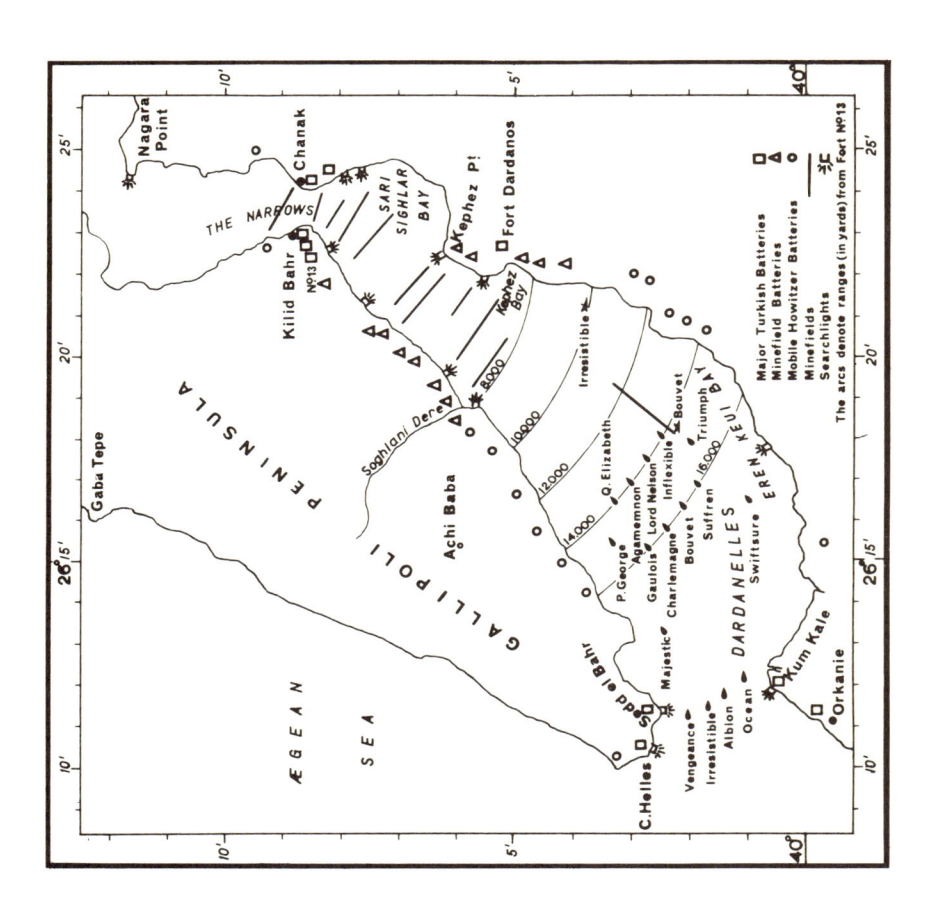

THE DARDANELLES
18 March 1915

addition to tests it was equally important to conserve flying hours for the all-important bombardment of the Narrows forts, especially in view of the many warnings of the limitations of the seaplanes. Machines were not getting off, or were unable to reach a satisfactory altitude, or were being forced to return with engine trouble.

But, as noted, there were no spotting tests to see what the seaplanes could do for the ships, and to insure that their spotting would be reliable by correcting any problems that might be uncovered. And instead of expending flying hours in essential work and reaching top efficiency in spotting, there were reconnaissance flights which contributed nothing to the success of the campaign. This policy culminated on 4 March, when, on the Admiral's instructions, the *Ark Royal* (she had moved up to the Dardanelles entrance) had aircraft flying for seven hours of precious flying time over the demolition parties, supported by marines, which had been put ashore after the outer forts had been silenced. It was "an unnecessary and useless job," Williamson claims. "The whole affair was a waste of time and effort," and the result was fiasco on the crucial day.[3]

That very evening (4 March), the *Ark Royal* received orders that an aircraft would be required to spot for the *Queen Elizabeth* the next morning, when she was to fire over the peninsula at the forts defending the Narrows. Williamson regards the mismanagement of the spotting for the *Queen Elizabeth* on 5 March as a crucial point — that, had, say, four aircraft been prepared days before and held in constant readiness "for the one job on which everything depended," anything would have been possible. The morning of the 5th was a disaster. The *Ark Royal* joined the *Queen Elizabeth* off Gaba Tepe. Lieutenant-Commander Williamson was to do the spotting, and his Captain told him to take any machine and any pilot he liked. He took the best one available; his own special machine had been hit and damaged on the "useless job" on the 4th. (Contrary to Admiralty regulations, on the way out from England he had altered this machine, exchanging the positions of pilot and observer to give the observer the best possible view for spotting.) Had they not "wasted" those seven hours the day before, he would have had his own good aircraft. At the appointed time on the 5th the aircraft was hoisted out. Williamson has recorded:

> It was a perfect day, with just the right amount of wind for taking off from the water, and we were soon in the air. It was an exhilarating moment. There below was the *Queen Elizabeth* with her eight 15-inch guns ready to fire and trained on the coast. The conditions were ideal; stationary ships and stationary target, only eight miles apart, and perfect visibility. I believed that there was every prospect of destroying the Forts, and that the Fleet would be able to go through the Straits and accomplish the object of the campaign by appearing off Constantinople. Few junior officers

have ever been in a position so favourable and of such importance, and I was thrilled with confident expectation. We soon reached 3,000 ft. and were ready to cross the peninsula to the target . . . Then it happened. In a moment the machine was out of control and we were hurtling towards the sea.

The propeller had broken up (the cause remains a mystery); the machine hit the water and practically disintegrated. Williamson and his pilot miraculously survived and were picked up by a destroyer. Another machine was sent up, but the pilot, unable to gain much altitude, was wounded by a rifle bullet and had to return. A third machine was able to signal only one spotting correction.

The negligible assistance of the seaplanes in spotting fall of shots was the immediate cause of the failure of the *Queen Elizabeth's* indirect fire at 14,000 yards range, since her gunlayers, unable to see the forts, were wholly dependent on aircraft for direction spotting.[4] Seventeen shells out of 33 fired landed inside the forts and did some damage, but no guns were hit. The fire had, however, taken the Turks by complete surprise and had, as a Turkish captain of artillery testified after the war, "a very great moral effect on gun crews as the batteries were quite unprotected from this direction."[5] This was not known to the fleet. On 6 March the indirect firing was continued, with the *Queen Elizabeth* having to move out to 20,000 yards from her objective and use full charges owing to fire from the *Barbarossa*. There was no seaplane spotting, only spotting by a battleship inside the Straits. It was ineffective and was abandoned owing to bad light.

"Had aeroplane observation been possible," the Mitchell Report concludes (speaking of the indirect bombardments of both 5 and 6 March), "there is little doubt that great damage would have been done to the forts, and that, with sufficient expenditure of ammunition, every gun might have been smashed. The forts were quite unprotected from this direction and each gun and mounting presented a maximum target. . . . Without aeroplane observation, little except moral effect could be expected, and this moral effect could be discounted unless the attack were accompanied by a simultaneous break-through of the Fleet."[6] The Report points out that it was not only on 5 and 6 March that the seaplanes "entirely failed to meet the main requirement of the Fleet, which was *accurate spotting*." Spotting results throughout had proved generally unsuccessful, and it was evident that *aeroplanes*, with trained observers for spotting, were a necessity. But these did not arrive until 24 March.

These comments need qualification. A popular view concerning the opening phase of the campaign (enshrined in the Mitchell Report) attributes much of the failure to the unreliability of the *Ark Royal* aircraft. This is misleading. The correct way to size up the situation is not by thinking of what might have been, but what actually was; not by what flying might have carried out, but by the number of hours actually flown; by the time aircraft were maintained in the air and made available for the Vice-Admiral to use as he wished. Between 17 February and 5 March, though not less than 50 hours were placed at his disposal, most, if not the whole, of the flying time had been expended in tasks none of which could have influenced the operation in any way.

In certain respects aeroplanes would have been more efficient than seaplanes for the work of the fleet. Their great advantages were that on several days they could fly from a land base when seaplanes were unable to rise from the sea, and they could fly higher and so out of rifle range. But it is not clear why aeroplanes should have been better for observing. In any case, they were not available, and one wonders, therefore, why the Vice-Admiral's staff had not given a modicum of thought to using the seaplanes available efficiently, instead of grieving all the time at not having aeroplanes. The same attitude in an extreme form appears in the Mitchell Report.

At the root of the failure to make the most of the seaplanes, according to Williamson, was the fact that no one in the fleet at the Dardanelles had any real appreciation of what these aircraft could do in the way of spotting. "But to be fair, there was no one *in the whole Navy* who had had any experience, and no doubt they would have learnt sooner, if we had not had such very bad luck."

A fully developed and reasonably competent Naval Staff would, even before the outbreak of the war, but certainly in its first months, have made a thorough investigation of methods of aircraft spotting, as well as of minesweeping and of the ammunition required to attack shore defenses successfully. The hard fact is that the fleet had received no special training in these matters. Perhaps this is too much to expect of a Staff which had only come into existence in 1912 and which shared the Navy's obsession with the big-gun duel between lines of dreadnoughts. In last analysis, however, good aerial spotting was not the crucial consideration.

By themselves the Narrows forts, owing to their obsolete guns and fire-control system, could not have prevented the fleet rushing the Straits. *The Turkish minefields had the most important function in the defense of the Dardanelles*: "to form an obstruction sufficiently formidable to prevent the rushing of the Straits." Responsible for the efficiency of the minefields was the able German Admiral Mertins, the chief technical adviser at the Straits. By the end of February 1915 there were five lines of mines across the Narrows and five more across the Straits just below Kephez Point, the southernmost, about 8,000 yards from the Narrows forts. Seventy-four guns, mostly 3-inch and 4-inch, and six 90-cm searchlights protected the minefields. The minefield batteries were partly fixed, like Dardanos, and partly mobile field artillery. Having no fire-control system, they were useless against fast-moving targets. N.I.D. had moderately accurate intelligence on these minefields. On 8 March, No. 11 line, "those mines of destiny," was laid in Eren Keui Bay. On this date there were 344 mines in place in the eleven lines. The mines themselves, mostly German contact mines, were reliable as regards holding their depth (about 14 ft.), and, though not of the most powerful type (the weight of explosive was 180-220 lbs.), they proved effective against the older battleships in the 18 March assault. The minefields were of such density (generally, spaced 44-55 yards apart, except for No. 11, where the mines were 110-165 yards apart) that the mathematical chances against effecting the passage to Nagara, beyond the Narrows, past the ten lines of mines, were better than 100 to 1; that is, not one of 100 ships could expect to reach Nagara. "It is, therefore, clear," remarks the Mitchell Report, "that the minefields did in reality constitute a formidable barrier against the Fleet and the above facts effectively dispose of the contention that the Straits might have been forced without sweeping the minefields."[7]

It was the original intention of the Admiralty that battleship fire should destroy the forts and silence the minefield batteries before an attempt to sweep the minefields. But this plan had been reversed when the battleship fire had proved ineffective. After the fleet bombardment of 25 February, which had reduced the outer defenses, the minesweepers were given the task of clearing the minefields by night sweeping. If successful, this would have permitted the battleships to destroy the forts at close range.

The minesweepers were a failure. On 10 March, at the seventh attempt, they actually reached the Kephez minefield (above line 8), but after one trawler had hit a mine and blown up, the others abruptly turned back. The following night they withdrew as soon as the first shells exploded nearby. Keyes, who was Chief of Staff to the Vice-Admiral, harshly criticizes the fishermen, whom he all but accuses of cowardice.

The sweepers faced certain handicaps which explain their miserable failure to do the job.

(1) The minefields and the associated gun and searchlight defenses appeared formidable, but the fire of the minefield batteries, even at short range, was more bark than bite — even the slow trawlers escaped, except on 13 March — but they did have a psychological effect on the trawler crews. Cruiser and destroyer attacks on the minefield batteries and searchlights during sweeping operations had little success, particularly since they could do nothing when the sweepers were actually in the minefield without adding to the dangers of the sweepers, owing to the difficulties of laying, ranging, and spotting at night. And yet no trawler was ever sunk by gunfire. The one badly hit had her engines stopped. The casualties were trivial compared to one attack on an enemy trench by the Army later on. How do we explain the timidity of the trawler's crews?

(2) The sweepers themselves were quite unsuitable for the task. There were 21 small North Sea fishing trawlers (17 or 18 were available on 18 March), but owing to repairs and rest leave to the personnel, the maximum operating at one time was seven. And their speed was a mere five knots when working in

formation, less the two-to-four-knot current of the Dardanelles down the Straits; with the current, seven to nine knots was possible. The slow speed prevented the sweepers from making much progress against the current with sweeps out. This necessitated sweeps being joined under fire and the mines dragged down the Straits. Moreover, the steel plating fitted to their bridges rendered the trawlers' compasses useless. There was no way to fix position and, even, at times of steering a course.

(3) The human material left something to be desired; raw fishermen from the northeast ports, trained in minesweeping, manned the trawlers. They were "Hostilities Only" ratings, mostly of poor physique and lacking in discipline. Their commander was a retired officer without sweeping experience. The draft of the trawlers, which was known to be greater than the depths of the mines, did nothing to improve the morale of the crews.

(4) The basic cause of the minesweeping failure was that the system of sweeping developed in British coastal waters was never intended for use in narrow waters under fire; it was indeed impracticable in such conditions — unless the minefield batteries were first mastered. Pairs of trawlers, 500 yards apart, towed a 2 1/2-inch sweep wire between and behind them. On catching a mooring wire, the trawlers towed it into shallow water for sinking by rifle fire. In the Dardanelles this procedure would have to be worked directly under Turkish guns, and for this the crews were not prepared. "The fact that their contract of service only insured them against the danger of exploding mines," Vice-Admiral K.G.B. Dewar has pointed out, "shows that exposure to shell fire was not even contemplated by experienced officers." There is indeed no earthly reason why they should have been expected to clear the minefields in the face of constant gunfire at short range. Nothing in their experience had prepared them for such an ordeal.

On 15 March Admiral Carden decided to revert to the original strategy of first destroying the forts with the battleships, then dominating the batteries, before attempting to sweep the minefields. But the next day he gave up his command for reasons of health. It was time he went. Carden was a charming man and an ideal peace-time society admiral, but he had none of the qualities needed for an admiral at war in the technical age.

Carden was succeeded by his Second-in-Command, Vice-Admiral J.M. de Robeck – a "fighting leader," an officer on his staff called him. "His character, personality and zeal inspired confidence in all," Churchill has remarked. But within a fortnight of De Robeck's appointment, Churchill was not to be numbered among his admirers. De Robeck's plan of attack called for a simultaneous "silencing" of the Narrows forts and the batteries protecting the Kephez minefield, so as to enable sweepers to clear a passage through the minefields.

Came the day of the attack, 18 March, which saw the heaviest and last naval bombardment of the forts. The results of the attack were:

(1) Fourteen old battleships (10 English, 4 French) and the *Queen Elizabeth* and the battle cruiser *Inflexible* attacked the Narrows forts, putting out of action four of the 19 heavy guns, though only temporarily.

(2) Four battleships attacked the minefield batteries, and, when the Turkish fire let up, six minesweepers were ordered up. They did not get within two miles of the Kephez minefield, still less commence to sweep it, before the gunfire sent them back, although none was hit. The minefields and their defenses remained practically intact.

(3) The loss through unswept mines of two old British battleships (*Irresistible, Ocean*) and severe damage to the *Inflexible*, as well as the loss of the old French battleship *Bouvet*, led to the abandonment of the attack. The basic cause of the failure on the 18th was the old story: the battleships needed to close the forts to silence them, but they could not do this until the mines had been swept; the mines could not be swept until the minefield defenses had been smashed — unless fast sweepers were used.

The particular villain in the piece on the 18th was line 11 of the minefields. Lieutenant-Colonel Gheel, a Turkish mining expert, noting that the enemy's ships inside the Straits at times manoeuvred in the slack water on the Asiatic side, off Eren Keui, thought it might be worthwhile to moor a line of mines there. Twenty were laid from the small minelayer *Nousret* in the night of 8 March in Eren Keui Bay, about 2 1/2 miles south of the permanent minefields and parallel to the Asiatic shore. Four of the twenty had been exploded in trawler sweeps on the 15th and 16th; yet on the morning of the 18th the commander of the minesweepers reported the area clear. It is a serious reflection on the competency of the officer in charge of the minesweepers that the 16 other mines were not discovered and swept before the 18th. It is also shocking that two pairs of trawlers exploded three of the Nousret mines on the 18th before they turned tail and fled. This was reported to De Robeck the *next day*! If reported at once, it might well have saved the three British ships that were to be mined in the same area (the *Bouvet* had already been mined).

Though little damage was done to the Narrows forts on the 18th (what there was was quickly repaired), the shortage of ammunition was of concern to the Turks. Having fired nearly 2,000 shells, the Narrows forts were down to 27 armor-piercing shell, the only kind effective against battleship armor. There was no shortage of ammunition for the medium and light guns. The morale of the defenders of the Straits, despite the ammunition shortage for the heavy guns, was high. There was confidence that their defenses would hold, were another naval attack mounted, and that their fleet would deal with any ships that did pass the Narrows. The view of the officers, Turkish and German, was not so sanguine, a number of them believing that a second naval attack would force the Straits or stand a good chance of doing so. The official Turkish view (1919) was that a naval "break through" was regarded as "not beyond expectation," and that the Government and public were in a state of excitement.[8] Neither De Robeck nor the Admiralty was aware of the state of Turkish morale, though they were partially aware of the ammunition factor.

"We are all getting ready for another *go*," De Robeck wrote to Sir Ian Hamilton (commanding the Mediterranean Expeditionary Force) on the 19th, "and not in the least beaten or down-hearted." The next day he informed Churchill, the First Lord, that he hoped to be able "to commence operations in three or four days . . . but delay is inevitable, as new crews and destroyers [being fitted as minesweepers] will need some preliminary practice."

De Robeck changed his mind on the 22nd. He had been brooding ever since the 18th on the loss of nearly a third of his capital ships, intensified by his remaining in the dark as to the precise cause. The only reasonable explanation of his losses seemed to be that the Turks were floating mines down with the current. Nobody seems to have realized that all the ships mined were out of the current. Had De Robeck shown more alertness, he would have had the positions of the mined ships plotted on his chart. The result would have indicated a line of mines, and a searching sweep by two "River"-class destroyers (six of them were being fitted with light sweeps as mine seekers) would have cleared up his gloomy broodings in half an hour. There was also the fear of possible minefields above the Narrows and about which they knew little, and of obstacles like large pontoons that might be sunk in the Narrows. As regards the first, there were no mines above the Narrows. Concerning the obstacles, Keyes has pointed out that such a blocking "would have been a physical impossibility" in the over 40 fathoms of water in the Narrows.

The "dominating consideration" in De Robeck's mind was his strong belief that if he got through into the Marmora without the destruction of the Narrows forts, that is, without the Army in control of the peninsula, his lines of communication would not be secure. This would place the fleet in a desperate position if its appearance off Constantinople did not quickly result in Turkish capitulation. And he did not expect that to happen. In his testimony to the Dardanelles Commission he maintained: "I think it was obvious [from the Turkish resistance on the 18th] then that the Turk was not going to give in easily; he was going to fight the whole way; and what one had been led to suppose, namely, that if we issued with the Fleet or arrived at Constantinople with the Fleet, there would be a change of Turkish Government, went by the board. It appeared clear that we had to fight the whole way and if we went to Constantinople we should have to go there with troops as well as ships."[9] What he had in mind by security of communications was the fear that, with the forts still intact, not many colliers and ammunition ships would be able to get through the Narrows, so that he would not be able to operate and maintain the fleet in the Marmora. It apparently had not occurred to De Robeck that he could have stayed off Constantinople for two or three weeks and then have returned, if no results were forthcoming.

The catalytic agent was Hamilton's assurance, at a conference in the *Queen Elizabeth* at Mudros on the morning of the 22nd, that the Army was ready to land at the toe of the peninsula, when it was ready, which would not be before 14 April. The object would be to occupy the Kilid Bahr Plateau and thereby dominate the forts on both sides of the Straits. Convinced that a combined operation would afford a better chance of success, the Admiral decided to drop the naval attack and to prepare the fleet for its venture with the Army. Keyes' estimate (22 March) that the destroyer sweepers would not be ready until 4 April clinched the shift in strategy for De Robeck: it was only a case of waiting another ten days for military support. As matters developed, the combined operation did not take place until 25 April.

The Admiral transmitted the vital decision to the First Lord after the conference in the flagship. He wired a full appreciation to the Admiralty on the 27th, which concluded: "With the Gallipoli Peninsula held by our Army, and a Squadron through Dardanelles, our success would be assured. The delay, possibly of a fortnight, will allow of co-operation which should really prove a factor that will reduce the length of time necessary to complete the campaign in Marmora and occupy Constantinople." A shocked and dismayed First Lord would have ordered De Robeck to make a fresh naval assault, but he was unable to carry the Admiralty War Staff Group, who would not overrule the decision taken by the naval and military commanders on the spot. He bowed to the decision of his professional advisers. "But with regret and anxiety," as he says.

Churchill, in the second volume of his *The World Crisis*, blasts De Robeck for his pusillanimity: he looked upon his old ships as "sacred." "The spectacle of this noble structure on which so many loyalties centred ... foundering miserably beneath the waves, appeared as an event shocking and unnatural in its character." Similarly, a quarter-century after the event: historical judgment would "pronounce him inadequate to the supreme moral and mental trial to which he was subjected."[10] De Robeck's decision required an extraordinary degree of moral courage. It was, moreover, based on the knowledge at his disposal. He was not aware of the shaky morale of the German and Turkish officers. He did not know what had brought on the losses of 18 March, which induced a more conservative strategy. Most importantly, as he had informed the Admiralty on 26 March, "To attack the Narrows now would be a mistake, as it would jeopardise the execution of a better and bigger scheme," that is, the capture of the Gallipoli Peninsula, in order to safeguard the fleet's communications once it was in the Marmora. *He could not have foreseen the failure of the Army.*

Keyes had, on 22 March, after the fateful decision in the *Queen Elizabeth*, pointed to the new sweeping force as the key to a successful fresh naval attack.

It was Keyes who saw that the revamped minesweeper organization had the power to drive a broad channel through the minefields and permit the fleet to pass the Narrows and anchor off Constantinople. The argument did not impress the Admiral. The new minesweeping organization would not be ready till 4 April, a mere ten days before the Army was prepared to land. There is also this hard-to-weigh factor. De Robeck, much as he liked and admired Keyes, doubted the soundness of his judgment. The Chief of Staff was, as Admiral J.H. Godfrey affirms, "a very gallant and inspiring leader but an indifferent chief of staff." He was never one to see beyond the immediate next step, and his judgment was clouded by his eagerness to get at the enemy. To sum up, there were good reasons, *in the context of the post-18 March period*, for De Robeck's preference for what appeared to be a surer and safer method of forcing the Straits. And yet ... and yet ...

The key to the situation was without doubt the minefields. It was obvious that the attacks on them had failed for three reasons: (1) the paucity of the sweepers; (2) the slow speed of the trawlers; (3) the inefficiency of the fishermen crews. Each shortcoming was remediable in fairly short order. As regards (1), in the sweeping operation in the night of 13-14 March the Turks had experienced great difficulty in controlling their guns on so many targets with so few searchlights. The minefield batteries lacked a proper fire-control system; theirs was barrage fire directed by telephone on a specified area. The logical deduction was that attacks by a large number of sweepers at one time might be successful. Concerning (2), the fact that the four to six destroyers which supported the trawlers in their sweeps were never hit by the batteries (their speed was their protection) pointed to the need for *fast* vessels, and of a draft less than the depth of the mines, able to sweep against the current, cutting the mine mooring wires as they proceeded. (3) was self-evident almost from the beginning and pointed to the need for a disciplined personnel. Substantial progress was made in all three directions within weeks. Indeed, most of the measures had been ordered as early as 14-17 March. Most importantly, eight "Beagle"-class destroyers were by 4 April fitted and trained as heavy minesweepers. They were joined by eight other "Beagles" on the 14th, and by eight fleet sweepers from the Grand Fleet (small steamers specially fitted as minesweepers) on the 16th. The last-named came out fully equipped and trained, and were manned with volunteers on arrival. By 18 April this new sweeping force was proven by trials to be capable of sweeping at 14 knots, at which speed they were probably immune to Turkish fire and at the same time were able with their heavy 2 1/2-inch clearing sweep to part the mine moorings when sweeping *up* the Straits, so avoiding having to tow the mines to a place of safety. The mines would now simply float away on the surface. And the draft of the destroyer minesweepers was only 10 1/2 feet, or well above the mines. The fleet sweepers, too, had a safe draft to go over the minefields. Even if not all of the second batch of "Beagles" were equipped and practiced by the 18th, there were more than enough trained and disciplined fast minesweepers by that date to sweep a channel a mile wide in one determined thrust ahead of the battleships. With, say, 10 "Beagles," there would be no fear of the sweep being ruined by a lucky shot or two knocking out a destroyer. The four fleet sweepers were better equipped to manage the surface sweep against floating mines, while they had a deep sweep out as well. They would have gone between the destroyers and the leading battleship.

Any possible doubts that the mine mooring wires of a line of mines would have been cut appeared to have been met the first time the destroyer minesweepers were used: in the combined operation of 25-27 April. For this operation, the orders read, "the efforts of the Navy will primarily be directed to landing the Army and supporting it till its position is secure, after which the Navy will attack the fortifications at the Narrows assisted by the Army."[11] In the first days of the operation the sweeping of the Dardanelles would be commenced by the destroyer minesweepers. (The fleet sweepers were not needed as sweepers and were used as troop carriers between the transports and the tows of boats off the beaches.) This facet of the operation called for progressive attacks on the minefield defenses from inside the Straits. These were to synchronize with the landing and capture of Achi Baba (first day), the advance of the Army on the Kilid Bahr Plateau (second day) and the occupation of the Kilid Bahr forts (third day). The Army landed on the first day but did not reach Achi Baba, then or later. Four pairs of destroyer-minesweepers swept an area two miles wide within 14,000 yards of the forts, including a line laid on 31 March (No. 13, off Tenker Dere). On the second day the destroyers continued sweeping to within 10,000 yards of the forts. On neither day were the destroyers hit by the Turkish guns, though under constant fire. On the third day the sweeping was continued up to 8,000 yards, including a line laid on 28 March (No. 12, off Chomak Dere), but this time one pair of destroyers suffered damage and casualties. "The sweepers were becoming more and more efficient and had by now demonstrated their ability to sweep minefields under fire. It was, however, considered that Phase II could be carried out in a shorter time and with fewer casualties when strongly supported by the Fleet. Sweeping operations in the Straits were, therefore, ordered to be abandoned until, by the advance of the Army, the Fleet could be freed to support the sweepers, which would also be reinforced by the remainder of the sweeping destroyers then supporting the Army."[12] The idea was to give the Army a chance to advance and so free the fleet to support the sweepers. "From the experience gained, however, there can be little doubt that the minefields up to 8,000 yards from the Narrows Forts could have been quickly cleared under the supporting fire of the Fleet."

Captain Boswell finds in the achievements of the destroyer sweepers on 25-27 April proof that a determined sudden thrust by perhaps 12 of the battleships, steaming at full speed (15 knots), following close behind the fast destroyer minesweepers, would have enabled them to go straight through to the Marmora at any time after 4 April.

Supposing the fleet at the Dardanelles had been led by a Nelson — or even by a Beatty! One can envisage him saying to the Generals at the fateful meeting in the *Queen Elizabeth* on 22 March: "As soon as my fast sweepers are ready, I am going through to Constantinople. If we don't then succeed in getting a peace settlement, it may be necessary for me to leave the Marmora and to land your army to occupy the peninsula, so that I can keep the fleet in the Marmora indefinitely." Further, let us suppose that the War Council, taking a real decision for once, had given orders to the Admiral to do it, accepting, if necessary, the loss of half the old battleships in view of the tremendous political prize. Had the fleet tried again, on 4 April, or, better still, 18 April, would it have got through? The crucial point is the effectiveness of the new minesweeper organization. One cannot prognosticate complete success by the minesweeper organization had a fresh naval attack been made. The destroyers had swept a large area in the lower Straits with complete success against the two new lines, 12 and 13, and with no losses; but these were short lines with few mines and they had not tackled the Kephez minefields, to say nothing of the minefield in the Narrows. One cannot therefore be certain that the reorganized force had overcome all the inherent defects in the old system. On the other hand, nothing is certain in war, and the performance of the destroyer sweepers was distinctly encouraging. Even had they failed on 18 April in the suggested operation, the whole naval attack could have been called off and the combined operation been permitted to proceed on the 25th.

Let us assume that the sweepers had preceded the battleships in a day-time attack (the fleet was not trained in night fighting, whereas the Turkish searchlights were quite well used) and had done their job, eliminating the minefield obstacle. What of the battleships? Four British pre-dreadnoughts and one French had reinforced the fleet, making it stronger than on the eve of the 18 March attack. Ample stocks of ammunition were on hand. It is exciting to think of the possibilities in the new sweeping force, and of the psychological and material effects of the fleet steaming at full speed straight at the Narrows — I do not think the Turkish gun crews would have stood up to the close-range rapid fire of the fleet: they had never experienced it — and the *Queen Elizabeth* and her indirect fire in support achieving the *moral* result needed. Aircraft and ship flank spotting would have put her salvoes on to the forts.[13] Let us also bear in mind that the ammunition situation at the forts was unchanged and the supply of mines limited. It does not require too much imagination to assume that the fleet would have passed the Narrows. There was nothing above Point Niagara beyond a couple of ancient batteries and the Turkish Fleet to stop it from arriving at Constantinople. De Robeck's instructions called for the destruction of the enemy fleet as "the first objective" after forcing the Dardanelles. This should *not* have posed any great difficulties.[14] When this had been achieved, the Turkish Army in Europe was to be cut off by the destruction of the Scutari-Ismid Railway and the Constantinople-Kuchuk Chekmeji road and railway. The next step would have been, with Russian naval cooperation, to force the Bosphorus. Finally, "Constantinople was to be summoned [to surrender] as soon as possible without prejudice" to these three objectives.

Supposing the fleet had anchored off the Turkish capital. Again, I believe there was at least a 50-50 chance of making the Turks ask for peace in April. The critics of another attempt to force the Narrows claim that the Turkish Government and military command would have moved out and declared Constantinople an open city — and then what? Churchill, Grey, Kitchener, and the Admiralty apparently thought that the arrival of the fleet off Constantinople in sufficient strength to defeat the Turco-German Fleet, if it had not yet been defeated, would produce decisive results by leading swiftly to a revolution against the Young Turks in power, withdrawal from the war by the new government, and the probable joining of the Balkan Powers in the war against the Central Powers.[15] The American Ambassador in Constantinople at the time offers solid evidence that the expectation of decisive results was well grounded. "Had the Allied fleets once passed the defences at the strait, the administration of the Young Turks would have come to a bloody end." Morgenthau claimed that the Ottoman State, which had "no solidly-established Government," was "on the brink of dissolution" on 18 March. "As for Constantinople, the populace there and the best elements among the Turks, far from opposing the arrival of the Allied fleet, would have welcomed it with joy. The Turks themselves were praying that the British and French would take their city, for this would relieve them of the controlling gang, emancipate them from the hated Germans, bring about peace, and end their miseries."[16] Morale in the capital had not improved noticeably by mid-April.

But how did the Turks and Germans visualize the scenario? In the judgment of the Turkish War Office, "It was impossible to estimate the situation which would have arisen if the Allied Fleet had forced their way past the forts, past the minefields, and entered the Sea of Marmora. However, if the British Fleet had attacked land transport from the direction of Bulair, and at the same time from the Gulf of Xeros [Saros], a very difficult situation would have undoubtedly arisen. It would have increased enormously the difficulty of transport between the Asiatic and European coasts and also in the Bosphorus and Marmora." If this attack by the fleet on Turkish communications from both sides of the peninsula continued, it was but a question of time before the Turkish Army had to capitulate.[17] Enver Bey (not to be confused with Enver Pasha, the Turkish strong man), Chief of Staff to Admiral von Souchon, Commander-in-Chief of the Turco-German Fleet, did not think the Allied fleet could get through the

Dardanelles because of the minefields (Souchon thought it would), but if it did, he believed that its arrival off Constantinople would have precipitated a revolution against Enver Pasha. Finally, we should bear in mind that the great boost to Turkish morale came when they realized that besides beating the fleet they had beaten the landings although greatly outnumbered. After that they were tough nuts to crack.

The political stakes were tremendous and well worth a fresh naval attack, even though half a dozen old battleships had been lost. There was, of course, no certainty of success, but the stakes and the not impossible odds would have justified a shot at it. There was none. Therein lies perhaps the greatest tragedy associated with the Dardanelles campaign: the lost opportunity of April, one of the most poignant might-have-beens of the First World War.

5 Smaller Navies and Disarmament

Sir Herbert Richmond's
Small Ship Theories
and the Development of
British Naval Policy in the 1920's

B. D. HUNT
The Royal Military College of Canada

On November 21st and 22nd, 1929, the London *Times* carried two lead articles entitled: "Smaller Navies – A Standard for All," and "The Capital Ship." But for the fact that they were written by Admiral Sir Herbert Richmond and released a few weeks prior to the opening of the 1930 London Naval Conference, they might have been seen as little more than the opening shots of another round of public debate on the subject of battleship versus aircraft and submarines – a question that had exercised naval thinkers' minds for years and would do so for many more to come. But these particular articles were more than the jottings of a proponent of the anti-battleship side of that well-worn argument. Written by a senior Naval officer who was recognized as a strategist and historian of consequence, they were nothing less than an open challenge to declared Admiralty policy and, indeed, to the very bases of British defence thinking in the post-war decade. Examination of the circumstances which prompted their publication offers some interesting insights into the assumptions underlying naval policy formulation and the realities of service politics in those years.

In strictly personal terms, the articles proved very costly for Richmond. He was reprimanded for publicly transgressing the official line, denied the senior posts for which he had always hoped and subsequently forced into early retirement. He was blamed, at least in part, for the outcome of the 1930 London talks which were seen by many as a total victory for the United States and the other naval Powers. A protesting Admiralty Board was forced to accept a further five-year stoppage in battleship construction and, more important a complete *volte-face* on the question of cruiser numbers, sizes and replacement schedules. This and the Global Tonnage figures so successfully opposed by the Admiralty at the abortive Geneva Conference of 1927 were then extended to destroyers as well. It was felt that Prime Minister Ramsay MacDonald's success in translating the Labour party's commitment to Anglo-American accord owed much to Richmond's public airing of his theories favouring small ships. "Unlike Mahan," Admiral Brian Schofield has recently written, "Richmond's theories made no impact on foreign governments, but caused considerable mischief with political thought in his own country...."[1] It has even been suggested that Richmond's arguments so impressed the Labour leadership that they seriously considered appointing him First Sea Lord. "He influenced Prime Minister MacDonald because the latter cared not a jot for defence, but did regard the budget and peace with high concern."[2]

Evidence to substantiate these contentions has not been found. Quite the contrary, Richmond's anti-battleship campaign of 1929 was a direct response to his awareness of how little real influence he wielded with either his political or his professional superiors. Moreover, his decision to publish his "heresies" was made in the full knowledge that it would cost him his career. In every sense, it was an act of conscience made necessary by the Admiralty's total inability to speak for itself and to protect what he passionately believed were the Nation's and the Navy's true interests.

Though directed to the immediate circumstances of the 1930 Naval Conference, Richmond's campaign was also the culmination of all his efforts to reform the Royal Navy from within; that is, to overcome limitations in mentality that accounted in large measure for the Navy's disappointing wartime performance and continued to underlay the myopic approach to policy development in the years that followed. He brought to bear the full weight of 45 years practical naval experience, his considerable influence as a writer and agitator, and his mature ideas as a strategist and historian. His action was therefore neither rash nor precipitate, but rather a calculated move by a man well-aware of his own value, and supremely confident about the soundness of his ideas.

Richmond's views on the ship size question developed over a long period beginning in 1920 in the months preceding the Washington Naval Conference. By that time, the Admiralty's seeming reluctance to enunciate broad policy lines for the post-war period had resulted in widespread public criticism and demands for open discussion of future ship-building programmes.

It was a difficult period in which a degree of uncertainty and extemporisation was inevitable. Four years of war had completely altered the international power structure. Extra-European interests assumed new, though undefined importance for Britain following the collapse of Germany and the rise of Japan and the United States. America's potential for outpacing Britain in naval construction alone demanded new assessments of the Anglo-American relationship. This diffusion in Britain's global defence responsibilities, when related to her losses in economic strength, also necessitated more clearly defined relations with the various Dominions, and for equilibrium closer to home. Successive British governments, all supporting the ideals of the League of Nations, the peaceful settlement of disputes and disarmament in specific areas, encouraged Europe's reconstruction as part of Britain's while avoiding wider commitments which would over-extend her own resources.

These considerations lay at the root of the infamous "Ten-Year Rule" which, reviewed annually, became the secret yardstick for calculating defence expenditure throughout most of the inter-war period. In politico-economic terms, a strong case could be made for the Rule's introduction; but in relation to sound defence policy formulation it was an insidious and unjustifiable rule of thumb. All too often it led the Service hierarchies to assume that its financial constraints absolved them from thinking too deeply about their most fundamental problems.

Technological advances further complicated the situation. The war had gradually revealed the possibilities of submarines, torpedoes, mines and aircraft, all of which challenged the continuing value of battleships which as the chief units of the Fleet constituted the index of British Naval supremacy. Not since the introduction of the *Dreadnought* had technological advance demanded such a fundamental rethink of basic naval doctrine.

The cumulative effect of all these factors was that Admiral David Beatty, as First Sea Lord from November 1919, was forced to play for time, to strike compromises between political economising and the preservation of as much naval strength as possible. The absence of a generally-accepted and coherent national strategic doctrine forced all three Services into an internecine struggle in which the object, apart from mere survival, was never very clear. Budgetary restrictions and the fluidity of the international situation abetted this tendency; but equally important were the attitudes of the men who held the senior positions in these years. In the same way that the Army was controlled by veterans of the Western Front, so the Admiralty was dominated by the Grand Fleet family — a family deeply divided within itself by the Battle of Jutland and its long aftermath of bitter controversy and personal recriminations. The "Jutland Controversy" of the 1920's, quite apart from its divisive and energy-diverting side-effects, induced a backward-looking mentality in senior naval circles which bore little relationship to post-war issues or the way they were being argued by the country's political leaders.

The effects of this Jutland-induced myopia can be detected at almost any level that the processes of British naval policy formulation are studied. The 1920's, Captain Stephen Roskill has suggested, were "a period of tactical sterility"[3] in which the primacy of the capital ship and the Battle Fleet concept dominated naval thinking to the extent that serious studies in other areas were neglected. Such studies as were carried out by the Naval Staff (notably, the Reconstruction Committee of 1918, and the Post-War Questions Committee of 1919),[4] the Staff Colleges and the Tactical School were pervaded by this obsession with Jutland and the lessons it appeared to offer. Restrictions on Fleet training and exercises severely limited opportunities for experimentation. In any case, with virtually every major Admiralty and Fleet appointment going to former Grand Fleet Officers, it was unlikely that radical or innovative thinking would be encouraged.

Richmond, as President of the Naval War College and as one of Beatty's unofficial advisors, was able to observe these developments with a well-informed though increasingly critical eye. Unlike his brilliant disciple, Captain (later, Vice-Admiral) Kenneth Dewar, who virtually destroyed his credibility by writing the Admiralty's *Naval Staff Appreciation of Jutland* (1922) — the centerpiece of the "controversy" — Richmond maintained his integrity and independence of Beatty. His friendship with Sir Julian Corbett, who was then completing the Jutland volume of the Official History,[5] undoubtedly permitted Richmond a clearer perspective on the issue than most people enjoyed. His Greenwich vantage point also opened a widening spectrum of contacts outside of the Navy. The fact is Richmond's horizon was expanding; these were the years of his maturation as an intellectual. Gradually, as his scholarly talents became more widely recognized, and as he turned his mind to the broader problems of national policy, his disillusionment with Beatty grew. This is not to suggest that Richmond was becoming a closeted academic undisturbed by current events. Quite the opposite, his attempts to influence the Beatty regime constitute some of the few instances where rationality, originality and a sense of deeper perspective entered post-war naval discussion.

His first opportunity came in late 1920 when he was invited to testify before the "Naval Shipbuilding Sub-Committee" of the Committee of Imperial Defence (the Bonar Law Enquiry) which was set up in response to a public outburst that attended the first indications that the new Naval Estimates would involve large additional expenditures. These included a resumption of battleship construction — eight super-Hoods, four to be laid down in 1921, and four more in 1922.[6] Firing the first salvo in a leader of November 29th, 1920, entitled, "The Navy: A Question for the Nation," *The Times* challenged the Admiralty to make known its plans and reasoning. Pessimistically, the Editor suggested that: "In the *arcana* of the Admiralty, plans may already exist for another revolution in naval shipbuilding as important as that effected by the production of the first Dreadnought. But we doubt it...." The flood of correspondence that followed, led by Naval Correspondents who knew something of the internal strife over Jutland, was too reminiscent of the pre-war Fisher-Beresford dispute to be written off as the squabblings of proponents of various weapons systems.

Significantly, these critics of the big battleships included several distinguished officers who had had much to say about their introduction.

Especially unnerving were the views of Sir Percy Scott. The much-venerated "Father of modern gunnery" in the Royal Navy had written only recently in his autobiography:

> Some officers say that the battleship is more alive than ever; others declare that the battleship is dead. I regarded the surface battleship as dead before the War, and I think her more dead now, if that is possible.[7]

Scott led the campaign in *The Times* to persuade the public that, in framing the new Naval Estimates, Beatty had not properly "analysed the basis of his convictions." The Admiralty's pre-occupation with World War I, the building of bigger ships and guns, and the entire system of professional preferment which was founded on battleship service, appeared to Scott as a mumification of the revolution he had fathered.

The outcry against the Admiralty's projected building programmes may have forced the Government into appointing the special C.I.D. Shipbuilding Committee, but it was a body whose purpose was anything but a searching and objective analysis of naval needs. Beatty was the only professional member. The others were Mr. Bonar Law (Chairman), Sir Robert Horne (President of the Board of Trade), and Churchill, Geddes and Long — the latter three all former or serving First Lords.[8] This plus the fact that Beatty made it abundantly clear to his Committee colleagues that the Admiralty's position on capital ships was firm, and that further studies would be pointless,[9] prejudiced all possibilities of any meaningful findings.

Altogether, fourteen witnesses were called. Of these, all but two — Rear-Admirals S.S. Hall (former Commodore of the Submarine Service, 1915-18) and C.M. de Bartolomé (former Third Sea Lord) — were Active List officers. The more senior of these were Grand Fleet officers and predictably pro-battleship in their thinking. Of the many retired officers and knowledgeable civilians whose names appeared in *The Times* only Admiral Hall was called to testify. Sir Percy Scott refused an invitation to appear arguing that it was a trap to muzzle his activities.[10]

Richmond had not participated in the public debate up to this point, although he was involved in the issue from early November 1920, when Beatty personally solicited his opinions. The First Sea Lord was clearly sensitive about the public attacks and was seeking arguments to bolster his position. With a characteristic regard for academic objectivity, Richmond noted in his diary; "I thought he was going about investigation the wrong way round. One should not try to prove what needed proving in one's own mind, but to find out what was right."[11] When he adopted this stance before the C.I.D. Enquiry he found himself uncomfortably straddling both schools of thought.

Like Hall and Bartolomé, Richmond had serious doubts about the continuing dominance of the battleship. But he would not go so far as to suggest that they should be replaced by less expensive submarines and aircraft, or that they were in any way rendered useless for the immediate future by these newer weapons. He agreed with the majority of the Active List witnesses that "the capital ship must remain, as it was in the recent war, the main support and cover for the lighter vessel by which sea-power is actively exercised both offensively and defensively...." But Richmond was alone in recommending that new construction should be postponed. More time was needed for experimentation and research. The uncertain international situation, new developments in forms of attack and defence, and finally, the nation's financial situation, convinced him that it would be folly to undertake costly building programmes at that time.[12]

The immediate effect of these arguments was to split the Enquiry down the middle. As a result, two Reports were submitted to the Cabinet. The first, signed by Bonar Law, Geddes and Horne, concluded that no adequate evidence had been heard to support contentions that the capital ship was obsolete. However, they added:

> ... although it may be true that in the naval warfare of the future, squadrons of such ships will play the same vital part as they have done in the past, the Committee feel it their duty to call attention to the evidence of Rear-Admirals Richmond and Bartolomé which points to the doubtful expediency of deciding to build big and costly vessels at the present time.[13]

The alternative Report, drafted by Beatty and signed by himself, Long and Churchill either ignored Richmond's testimony or reworded it so as to de-emphasize its significance or discredit its author.[14]

The Cabinet was then faced with two conflicting assessments. From the Admiralty's point of view this was highly disappointing; the case against the battleship had not been made, but neither had the Enquiry justified, in completely unequivocal terms, the need for an immediate and extensive programme of building new ones. By the time of the Washington Conference, firm decisions in this respect had still not been made. Cabinet approval was given for the Admiralty to complete its Super-Hood designs, and orders for two were placed in late October 1921. But, within a month, work was suspended, and then in February 1922, cancelled altogether. By that time, events in Washington had altered the question entirely.

The net effect of Richmond's testimony before the C.I.D. Enquiry on these later developments is difficult to assess. Together with Bartolomé and Hall, he had at least ensured that something more than a whitewash of Beatty got to the Cabinet. Clearly, the doubts he expressed played into the hands of those Cabinet members who were already concerned about increased naval spending and its effects on Anglo-U.S. relations. Richmond's views probably reached the Cabinet by even more direct means than the C.I.D. Report. Hankey had written in December 1920, in a "purely private and personal capacity," asking Richmond for a brief on capital ships which he could show "in strict confidence, if you think fit, to [the] Prime Minister, Mr. Bonar Law, and Mr. Balfour." [15] Whether Richmond replied is not known, but it was the kind of opportunity he seldom ignored. His correspondence of this period shows that he was anxious to precipitate a full-scale re-evaluation of naval policy in which the influence of the battleship enthusiasts would be placed in reasonable perspective. Had such a study been undertaken, the outcome of the Washington Naval Agreements may well have been fundamentally altered inasmuch as the British Naval delegates would not have been caught so flatfooted as they were.

Throughout the talks Beatty and his advisors found themselves fighting a defensive rear-guard action instead of a well-directed advance along lines predetermined by thorough assessments of what was desirable and possible. The Admiralty's approach to the Conference was generally open and flexible, and was guided by two general principles; first, that any agreements reached should not jeopardize Empire naval security; and second, that in view of high public expectations, there should be nothing in the British position for which the Admiralty could be held accountable should the talks fail. Beyond that, the Admiralty followed an essentially opportunist approach of "no concrete proposals." The British naval delegates were directed to confine themselves to considering proposals put forward by the other delegations, if possible guiding discussion into channels which "the Naval Staff consider the most likely to be productive of a practical solution." On one point only was the Admiralty firm; there would be no extension of the unofficial building holiday which had been in force since the end of the war. [16]

Secretary of State Hughes' opening remarks to the Conference on November 12th, 1921, thus took everyone by surprise. Especially disturbing were his proposals for abandonment of all capital ship construction, including the scrapping of Britain's Super-Hoods which had only just been ordered. Suggestions that cutbacks might include post-Jutland ships had simply not been anticipated by the British: "no future historian," wrote one observer of that opening session, "will ever find anything so vivid as that look on Lord Beatty's face as he listened to Mr. Hughes' announcement."[17] The Americans had gained a tactical surprise from which the British naval delegates never fully recovered. Subsequent acceptance of a ten-year building holiday, the system of determining numbers of capital ships to be retained under the 5:5:3 ratios on a Total

Tonnage basis, and the restriction of individual battleship displacement to 35,000 tons, all ran counter to professional naval thinking and owed more to the political assumptions of the so-called "new diplomacy" than to strategic, tactical, or technological considerations.

It was the essential artificiality of these criteria which moved Richmond to re-open public debate at this point by asking why even lower maximums were not considered. Using the psuedonym *Admiral*, he inquired in *The Times* of November 23rd, 1921: "Why 35,000 tons? What is there in this number of importance? Why not 34,000 or 30,000 or 20,000 or 10,000?" There was no military justification for the 35,000 ton limit. Arguments favouring larger ships on the basis of protective and striking power, the need to produce vessels more powerful than those of an enemy, or the need to keep armour-plate firms in existence, he dismissed as mechanical, not military reasons.

> Now that our statesmen are sitting round a table and discussing this in friendly fashion they have such an opportunity as has never occurred before. . . .
>
> The sole qualifications of a ship of war are that she shall be able to go to sea, and fight. There is a limit beyond which it is quite unnecessary for her to go. I have suggested 10,000 tons but this is guesswork. It may be 6,000 or 10,000. I am sure it is not more.

So far as can be determined, *Admiral's* identity was never discovered by the Admiralty. Eight years later, however, almost to the day, his repetition of this stand would be the immediate cause of his release from the Navy. But in 1921, neither his action nor his ideas had any appreciable effect on events in Washington.

The Cabinet did discuss Richmond's article,[18] and directed the Admiralty to comment on "the truth" of his contentions that 10,000-ton ships were equal to larger vessels in protecting Britain's interests. Paraphrasing Richmond's *Times* letter, Lloyd George put the question to the Admiralty:

> If it is held that there is enough in the above considerations to justify our advocating a low tonnage limit, it may very likely recommend itself to the American Government, since it will go far to solve the problems of rapid replacement, presented by the ten-years naval holiday.
>
> . . . If France and ourselves supported it together the appeal to American sentiment morally, as a great further limitation of armaments, and materially, as a higher promise of solvency on the part of their debtors, would be irresistible.[19]

The Admiralty was predictably unenthusiastic and made no attempt to meet Richmond's contentions head on.[20] As a result they never came before the delegates at Washington. In any case, it was too late. Negotiations were well-advanced and a radical change in approach at that point might well have produced chaos. The Admiralty's position had been steadily undermined as it was; a change to a new line to which no thought had been given would have weakened its position further. Had these ideas been thoroughly examined by earlier investigative bodies (namely, the Reconstruction, Post-War Questions and the Bonar Law Committees), it is possible that more room for manoeuvre would have been available. Though his ideas came too late to affect the Washington Agreements, they remained for Richmond a subject for continuing study and reflection.

Over the next few years, Richmond tested and sharpened his thinking against the criticisms of various friends, and his War Course and Imperial Defence College students. If his experience of 1920-21, particularly Churchill's violent attack on his testimony before the Bonar Law Enquiry, had taught him anything, it was the need for greater precision in his thinking, though even that would not guarantee eventual understanding or acceptance of his ideas. In May 1926, after lecturing at the Royal Institute of International Affairs on the "Limitations of Armaments," he was politely but firmly reminded by the Deputy Chief of the Naval Staff (Vice-Admiral Sir Frederick Field) that in expressing the opinion "that we could reduce naval armaments by a further limitation in the size of ships," he had transgressed the official line. "I feel sure," the unofficial reprimand concluded, "you will refrain from giving public expression to any views which would embarrass the Admiralty in connection with this question."[21] Even Chatham House meetings, which were considered strictly private, offered no protection for a heretic in this area. Richmond was therefore fully alive to the dangers involved when, in the summer of 1929, he mounted a concerted attack on the Big Ship advocates and the possibility of another misdirected disarmament conference.

His first step was to release in the August number of the *Naval Review* a series of articles entitled, "What is it that Dictates the Size of the Fighting Ship?" These, together with supporting articles and letters from supporters at the Staff College, the Naval War College, the Imperial Defence College, and from various prominent personalities in the publishing world, he planned to release as a direct provocation to his professional superiors. As he confided to retired Admiral W.H. Henderson, Editor of the *Naval Review*:

> The Admiralty are now going to be put on their mettles; they will have to shew cause for their adherence to large ships and I should not be surprised if this Labour Govt. [sic] did not shew a desire to

take the lead, and propose something positive in economy, instead of being dragged at the heels of America all the time.

He welcomed the thought of a flood of replies:

> The more I am attacked, the more openings I have with dealing with sophistries. . . .[22]

His message was simple. He wanted to discover the primary determinant of size: "What is the criterion? Is it not possible to relate the size of the ship to something absolute?" Range, endurance, speed, gunpower and protection were all relative considerations. The then accepted criteria, he suggested, were not related to the positive requirements of what a ship had to do, but rather to negative considerations of what she must survive. Defence, he observed, was the ruling factor, rather than function. Since fighting ships were the units which constitute a Navy, their function was directly related to the way naval power is exercised, or more concisely, the object of naval warfare. And that, in his view, was the control of maritime commerce. The employment of naval forces as battle fleets, escorts, carrying forces, blockading forces, raiders, etc., were matters of strategy and tactics – the *means* of achieving the prime object. Thus Richmond advanced his theory that the size of the individual fighting ship was in the final analysis, determined by the strength of the merchant vessel, "whose arrest is the final object of war at sea."

There was nothing, he reasoned, in recent technological developments which inherently justified size increases. If anything, the War had shown that the smaller units were the least vulnerable and that they possessed most of the offensive threat to the new weapons. Reduced size would also decrease docking and repair costs. Most important, size reductions would open the way for simpler disarmament formulae by removing the classifications in size, calibres and numbers which had been the principal areas of international disagreement. Specifically, he envisioned his ideal capital ship as having the following characteristics: a displacement of 7,400 tons, endurance of 7,000 miles, a maximum speed of 28 knots, and an armament of eight 6-in. guns.

The replies to Richmond's proposals were generally disappointing. None expressed serious disagreement with his logic. Most simply reflected their authors' discomfort in working as Richmond had at a purely theoretical level. Not one advanced arguments which he had not anticipated. His lengthy and tightly reasoned responses to each in turn are interesting inasmuch as they illustrate Richmond's untiring patience when it came to enlightening interested correspondents. But they also help to demonstrate the simple, yet compelling logic of his small ship theories. The difficulty was one of how discussion should proceed if, in his phraseology, it was to be "real discussion, discussion on scientific lines."[23]

Richmond approached the problem on several levels, separately and successively, beginning first with the abstract theory "unaffected by national advantages and disadvantages" which necessitated a particular size; that is the strategic and tactical principles which governed absolute size. Next, he would examine the advantages and disadvantages of reducing sizes to this theoretical limit. Finally, he would study the possibilities of international agreements along these lines. His *Naval Review* articles were trial balloons related only to the first level. What he sought up to that point was a consensus on his theoretical conclusions. Once their validity was acknowledged, the other two stages would follow.

The replies he received convinced him that he was right. This being so, he confided to Admiral Roger Backhouse,[24] the Third Sea Lord:

> ... it does not become easy for individual nations to oppose a scheme which is logically correct, which means a vast saving of money, which tends to reduce the jealousies now existing, merely on the basis of some supposed national requirements. If, indeed, national requirements are adversely affected, what this would shew would be that the theory was wrong.[25]

The Americans would be obliged to justify their faith in large battleships and 10,000-ton cruisers, and to challenge Richmond's premises. "Let them disprove them," he continued, "I think they would find it difficult, if as you say, the arguments are logical." The American sailor would be hard pressed

> ... to admit that *he* [sic] must have an 8" gun to fight a wretched merchantman: that he must have 35 knots to catch a 15 knot ship ... The fact is that we have shewn up the nonsense they talk: and we ought to do so. Further, it would be awkward for the Americans who cry out that their purpose is idealistic, that they desire to reduce the burden of armaments, to reject proposals which not only materially reduce the cost, but enable them, at a reduced cost, to obtain that 'parity' which they affect to believe is necessary for their security.

Expectation of determined American opposition was not sufficient grounds for British hesitation. On the contrary, he argued, it was the very reason Britain should take the initiative.

> Are we to be deterred from making suggestions which we believe to be logical and sound, to injure no one, to profit all . . . I say the sound policy is to put them forward and make the American come into the open and disclose whether all his talk is hypocrisy; and whether his intentions are genuinely to benefit the world or merely to increase his own prestige and strength. I would (having satisfied ourselves, of course, as to what we think) throw this bombshell into their camp, and confront them with a proposal which they will find it difficult to reject because they will have to deal with public opinion. Such a proposal coming from the Admiralty would have an uncommonly good effect. No one, alas, can hide from himself that the Admiralty has not got what is called a 'Good Press.' The silly picture — caricature — of the Admiralty as a body of old red-faced, white whiskered gentlemen with large bellies, crying out for dominance of the world at sea, is really believed by the foolish people who rule us from GRUB Street. Because of this, the fantastic nonsense talked by the Air people gets support. They will not believe that ships are more important than aircraft. They do not want to believe it. But give them evidence that the Admiralty really means to make a drastic cut in expenditure — and it must be something big and unexpected — confidence would be assured, and with it the support of the country.

Backhouse was clearly impressed with Richmond's reasoning, as indeed were a number of distinguished Flag Officers on the active list.[26] Others, though they could not embrace all of Richmond's ideas, were reported to be anxious to reverse tendencies in building and design policy which had been set in train by Fisher 25 years before, but had their reasons for not wanting to make their views known. For one thing, radical change in Admiralty policy would mean public admission of errors in the past. Moreover, it was known that the First Sea Lord, Admiral Sir Charles Madden, was to be succeeded by another Fisher disciple, and this implied "a compliance on the part of others." Backhouse was reported to have admitted that the "Juggernaut at the Admiralty, manned by 'Materialists' almost exclusively, is too powerful for him to stop. He, and other officers are convinced that the reform of our terrible war and post-war errors of policy must come from outside. . . ."[27]

This was the responsibility Richmond accepted when he submitted his *Times* letters of November 1929, challenging the Admiralty, Government and public alike to speak out on the vital issues which the London Naval Conference was about to deliberate. The following excerpt from a letter to Admiral Henderson leaves no doubt as to Richmond's motives:

I cannot help it if the Admiralty object; or refrain from doing this. I am so convinced that it is for the good of the nation that this matter should be considered, when it comes to be considered next January, on some higher plane of policy, strategy, and logic than it was at that unfortunate Conference at Washington, that whatever my personal advantages may be [sic] — or rather, disadvantages — do not matter. My professional aspects are not to be considered. Perhaps I exaggerate the importance of my views. I may. But I hold them very strongly, as the result of long thought — eight years and more — and much discussion. I have invited and received criticism freely and I doubt, I hope not in consequence of any vanity, that there is anyone who has applied himself more to the study of the subject than I. That does not mean that I think I must be right. I believe I am, because no proof has been given me that I am wrong. What I propose is, I know, difficult to put into Execution [sic] under the conditions of Timidity [sic] that govern our policy and practice. But I am sure that a few strong men could do it. Whether such exists remains to be seen.[28]

Stated simply, Richmond set out to demonstrate that previous international limitations agreements had been political compromises based on mathematical calculations believed to balance off the "supposed interests" of the various Powers. With the London Conference only two months off, he called for an agreement based instead on principles. Surveying the functions for which navies are maintained, and suggesting the varying degrees of importance which individual nations attached to them, he concluded that the only "true criterion" of quantitative naval limitation should be the strength of the weakest naval power — the one to whom naval defence was least vital. This much established, the greater naval Powers could then fix their requirements thereby satisfying the twin needs of security and economy.

Richmond called for a more rational approach to disarmament than that of simply fixing for individual countries the number of ships of defined categories, and the numbers and sizes of weapons carried. He rejected the system of complicated regulations that had stifled thought and been subject to abuse. He sought agreements which would allow each Power to fix, in concert with the others, its total requirements basing them on the weakest Power criterion as the minimum measure. Each nation could then build whatever types it preferred as long as they did not exceed the agreed-upon adequate tonnage limit for a capital ship and that, in his view, should be very much less than had been accepted over the preceding decade. In this case, he suggested 10,000 tons. The key assumption of his "smaller navies" proposal was, of course, that international agreement on a tonnage limit for capital ships was possible. As the Editor of *The*

Times noted, not everyone would accept Richmond's premises or his conclusions. However, he suggested:

> The views which he now expresses on what is perhaps the most important question of the day can hardly fail to make for a clearer understanding of the problems which will be set for solution in January.[29]

The Admiralty, however, was not so open-minded. When questioned in the Commons as to whether the articles written by "an admiral on the active list" had been seen by the Admiralty, and whether they represented the official view, Mr. Alexander the First Lord, emphatically denied both Richmond and his ideas.[30] Within a week, Richmond received an unpleasant official reminder about the provisions of King's Regulations, Admiralty Instructions, and the Official Secrets Act concerning public expression of views on controversial questions.[31] This was not surprising. The public reaction, however, was.

To have attacked the *status quo* and received the almost complete support of the London daily press was no minor accomplishment. Even those commentators who challenged Richmond's specific conclusions, admitted the timeliness and the merit of his ideas, and indeed, the personal implications of his action. As *The Spectator* noted:

> No one could give intelligent attention to Admiral Sir Herbert Richmond's recent articles ... without having his faith in the big ship as an essential feature in a modern fleet gravely shaken if not dispelled. The great merit of such articles is that they bring the battleship to the bar and throw on its defenders, the onus of establishing their case....[32]

The Editor of *Foreign Affairs* agreed. Referring to Richmond as "the ablest mind in the navy," he considered his case against the big ships to be "overwhelming."[33]

But expressions of sympathy from the Press and other supporters did not mean that Richmond's ideas necessarily affected the outcome of the London Conference. Admiral Chatfield's bitter contention[34] that Ramsay MacDonald and the Labour Government were heavily influenced by Richmond's theories is highly doubtful. There is nothing in Richmond's papers to indicate that he made personal contact with the Prime Minister on this question. Moreover, the Agreements themselves bear little resemblance to Richmond's proposals. The fact is, MacDonald's broad policy lines were well-formulated before Richmond went to *The Times*. Indeed, as early as August 1929, MacDonald had made known his Government's willingness to compromise the Admiralty's longstanding claim for seventy cruisers and to reduce that figure to fifty. By year's end, it was public knowledge that the Conference would also seek agreements to further postpone battleship construction on the assumption that the Americans would not raise the ever-contentious "Freedom of the Seas" issue. The net effect was that the London Conference extended the earlier Washington agreements, the original drawbacks of which reappeared in aggravated form. Battleships were not to be replaced until after 1936, but the vessels themselves were not called into question. The Americans reciprocated Britain's submission to cruiser parity by agreeing to reduce their numbers of 10,000-ton Washington Treaty cruisers from 24 to 18. In view of France's and Italy's refusals to be bound by any of these limitations, the London Naval Agreements had serious implications which MacDonald and President Hoover seemed determined to ignore in the name of trans-Atlantic tranquility.

Richmond was utterly disappointed with the results. He expressed hope that improved relations with the Americans might outweight some of the disadvantages. He also saw in the postponement of capital ship construction some promise that they might eventually disappear altogether. But there was very little else to commend the Agreements in his view. He lamented the fact that his small-ship theories had not been taken seriously: "an unparalled opportunity" for significantly reducing expenditures had been missed. He especially regretted Labour's concessions on cruisers. For years afterwards, he maintained that this decision and the standards established in 1930 were the major cause of the Royal Navy's critical cruiser, destroyer and flotilla problems in World War II. More than anything, he regretted the application of "arbitrary and mathematical bounds" for fixing the size of navies. The ratio system, the "doctrine of material parity," had been useful at Washington as a first step, "a clod to stop a leak in time." But, he suggested:

> ...the merit of this formula, its simplicity and mathematical conclusiveness, has tended to impress itself not only upon all our subsequent negotiations but upon our very way of thinking. The only meaning which the phrase 'naval parity' conveys to most people at the present time is this equation of weight and strength of armament . . . mere mathematical equation in naval tonnage brings only a specious resemblance of naval parity unless it achieves those aims which are the true pacific purpose of the whole negotiation.[35]

In the end, Richmond failed to re-direct discussion along more meaningful lines. His espousal of a "qualitative" approach to disarmament elicited from the Admiralty only the negative response of ridding the Service of a heretic who for too many years had irritated naval officialdom. That no attempt was ever made to examine the validity of Richmond's small ship theories, and that Madden never once admitted that *The Times* articles were the real reason for his subsequent removal, suggests how dimly Richmond's purpose was appreciated. Following his retirement, Richmond continued to question the adequacy of cruisers and destroyers for British needs and challenge the Admiralty's persistant faith in heavy battleships. In 1936, for example, when he was called before yet another C.I.D. Enquiry — this one to study the capital ship's vulnerability to air attack — he found it necessary to emphasize once again the futility of investigations so restricted in assumptions and viewpoint. The question of whether a battleship could be made safe against bombs was meaningless unless it was related to more far-reaching studies of how fleets would be used — a problem of strategic philosophy that transcended in importance mere questions of technology.[36]

By way of conclusion, it may be of interest to note how Richmond's concept of "qualitative" disarmament influenced the thinking of at least one other important defence intellectual of the period; namely Captain Basil Liddell Hart, whom he came to know well in 1929. Prior to 1930, Liddell Hart's writings had been devoted almost entirely to problems of battlefield mobility and the development of armoured forces and tank doctrine. In 1931, he turned to two new themes. The first was his esposal of a maritime strategic or "limited liability" outlook which he first expressed in a R.U.S.I. lecture in January 1931, and later expanded in his highly successful book, *The British Way in Warfare*. Richmond was the only one in that R.U.S.I. audience to endorse his arguments.[37] Since they amounted to a recapitulation of many of his own conclusions, this is hardly surprising. The other question to which Liddell Hart had not given serious thought was disarmament. In May, 1931, he was asked by Sir Samuel Hoare to study the question and to act as an unofficial consultant to the Government on the issue as it developed in the weeks preceding the 1932 Geneva Conference.

Like Richmond, Liddell Hart came to the conclusion that previous attempts at general disarmament had failed because they were based on numerical standards. Seeking a radical solution that would minimize bickerings over ratios and numbers, he saw that the problem was essentially a qualitative one. Specifically, it boiled down to limiting weapons, such as heavy guns and tanks, which inherently favour the offensive. Abolition of these weapons would limit the chances of successful aggression, leaving countries free to ensure their own sense of security by building whatever types of defensive systems they desired. Referring to this whole mental process, Liddell Hart later recalled:

> It was the hardest test that had ever confronted me in striving to take a completely objective view ... For I soon realized that the obvious solution would entail annulling not only the development of tanks as a military tool but the whole concept of reviving the power of the offensive and the art of war, by 'lightning' strokes with

highly mobile mechanised forces — thus cancelling out all I had done during the past ten years to develop and preach this new concept. If I propounded the 'disarmament' antidote to it, and helped to obtain its adoption at the coming conference, it would mean strangling my own *baby*.[38]

Yet, this in effect is what he did. His plan for an "offensive-curbing method of disarmament" was favourably received by the British Cabinet and later, at Geneva, made a considerable impact on a number of delegates to the Conference. But attempts to introduce it as a basis for concrete agreements were ultimately rendered abortive by the unexpected march of international events which wrecked the entire Conference.

These considerations go beyond the immediate context of this paper; but they do indicate that Richmond's disarmament ideas of 1920-30 survived his enforced retirement from the Navy. The similarities of Liddell Hart's "qualitative" approach are sufficiently striking to suggest that Richmond was the major moving force behind them. Sir Basil specifically denied this in 1969.[39] Nevertheless, it is clear that the two men did exchange views, although the exact nature and scope of their contact remains obscure. This is partly because much of their contact was personal and verbal. The surviving documentary evidence does show, however, that the two men were in frequent sympathetic consultation and that Richmond's ideas were well-known by Liddell Hart. By late 1930, Richmond's *Economy and Naval Security* (London, 1931) was well-advanced; between October 1930 and January 1931,[40] they discussed all its major themes.

6 Canadian Maritime Strategy in the Seventies

G. R. LINDSEY*
Defence Research Analysis Establishment, Ottawa

Compared to the problems of British naval strategy or German navalism prior to World War II, or to American naval strategy subsequent to World War II, Canadian maritime strategy in the 1970s is a very minor and unimportant subject. It is unlikely to have much significance on the world stage whether it is conducted well or badly. But it could have very considerable significance for Canadians, and just possibly for all North Americans.

Those professionally concerned with the planning of Canada's maritime strategy are looking for all the help they can get from whatever source may be of use. I think it quite possible that history can be of considerable use. But it could be history of subjects that do not first suggest themselves in the context of naval strategy — perhaps the history of mankind's search and conflict for sources of food and raw materials, or his quarrels over national boundaries, rather than the selection of the maximum tonnage of battleships. Or perhaps the history of the means by which national policies are altered under the British system of parliamentary government with an apolitical public service, rather than a blanket condemnation of militarism or of wooden-headed admirals.

And it occurred to me during our extremely interesting discussions that, if we are still discovering new material and struggling to resolve conflicting

---

* The contents of this paper are the responsibility of the author, and do not represent the opinion of the Defence Research Board or the policy of the Department of National Defence.

interpretations of events that occurred fifty years ago, we could afford a little charity toward the poor officers, officials and politicians who are struggling to predict developments that have not yet happened at all.

Naval Developments Prior to the Seventies

Before discussing Canadian maritime strategy for the nineteen seventies, it is probably desirable to spend a few minutes recalling some of the main changes of the forties, fifties and sixties.

The main Canadian maritime role during World War II was antisubmarine protection of the North Atlantic convoys. The Royal Canadian Navy manned and operated a large number of small escort vessels[1], and the Royal Canadian Air Force flew maritime patrols. Canadian maritime forces also fought in the English Channel, the Mediterranean and the Pacific, but the most important contribution was to the Battle of the Atlantic.

Between 1946 and 1955, the Soviet Union built up the largest fleet of attack submarines ever seen. The member countries of the Atlantic Alliance prepared for another Battle of the Atlantic, equipped themselves with escort destroyers, maritime patrol aircraft and escort carriers, and made plans for the control of merchant shipping and the sailing of convoys. The Canadian role was escort of convoys, for which we had an escort carrier,[2] destroyers and frigates. Our ships did not have sufficient speed to escort fast carrier strike groups.

In 1950 the RCN had ships in the Korean theatre six months before Canadian ground troops arrived, and maintained three destroyers throughout the campaign. The versatile capabilities of the force are well described in a paragraph from the Naval Historical Section:[3]

> For over three years these hard-working little ships joined their colleagues in the United Nations force and the ROK Navy in performing a great variety of tasks: maintaining a blockade of the enemy coast; protecting the friendly islands on both coasts from amphibious assaults and sneak raids; providing support for the coastal flanks of the United Nations armies; bombarding Communist installations, gun emplacements, troop concentrations and road and rail lines along both the east and west coasts; screening the United Nations carriers from the ever present threat of submarine and aerial attack; supporting the numerous friendly guerillas and ROK regulars in their unremitting harassment of the enemy mainland and islands; bringing aid and comfort to the sick and needy of South Korea's isolated fishing villages; and performing the countless other tasks that fell to the lot of the UN destroyers serving in the waters around Korea.

During the interval from 1956 to 1963 doubts began to be felt concerning the plausibility of a long "broken back" war of attrition, in which the success of the NATO forces would depend on supplies fought across the Atlantic over a long period. It was not believed that the devastation of nuclear war could be endured for long, or that seaports and inland communication would remain able to move supplies even if the ships succeeded in reaching land. Also, the technical problems of convoy protection became more severe when they were threatened by nuclear weapons, delivered by air-to-surface missiles or ship-to-ship missiles. In order to prevent the loss of several ships to one weapon, it became necessary to increase the spacing between adjacent ships. This extended the perimeter of the convoy and made it easier for a submarine to penetrate the protecting screen unless the number of escort ships was greatly increased. And, to add to the vulnerability, the presence of a convoy at sea and its precise location were likely to be discovered by reconnaissance aircraft or satellites. The chief advance in antisubmarine technology came with improved sonar, including substantial advances in sonobuoys which enabled an aircraft to detect a submerged submarine by acoustical means.

In 1956 the aircraft carrier *HMCS Magnificent* helped to transport the Canadian contingent to the United Nations Expeditionary Force in Egypt.

Between 1964 and the present day the probability that it would be necessary to protect large merchant convoys appeared to be further reduced. Nuclear powered submarines appeared in large numbers, faster than the surface ships designed to fight them and not requiring to come to the surface for weeks on end. In addition to the usual anti-ship torpedoes, some Soviet submarines were armed with surface-to-surface missiles able to engage ships or land targets. But most important of all were the ballistic missile firing submarines, which soon replaced the strike carriers as the main maritime weapons for strategic deterrence.

Ship-based antisubmarine weapons were given greatly extended range by the use of rockets to propel torpedoes through the air to the vicinity of the target, after which they entered the water and homed on their target. And an even better weapon was the destroyer-borne antisubmarine helicopter, equipped with dipping sonar for detection and torpedoes for attack. Sonar mounted on the ship's hull was supplemented by variable depth sonar, towed behind a destroyer at the best depth for the water conditions.

The chief quarry of the Canadian antisubmarine forces now became the ballistic missile submarine instead of the attack submarine, and to an increasing extent the quarry was propelled by nuclear rather than diesel-electric engines.

In 1964 the Canadian carrier *HMCS Bonaventure* helped to transport heavy equipment for the United Nations force in Cyprus, but in 1971 she was withdrawn from service. Three operational support ships were acquired, and three O-class diesel attack submarines. About half of the destroyers were converted to carry antisubmarine helicopters.

Necessary Areas of Canadian Maritime Activity

The Defence White Paper of 1971[4] defined four major areas of activity for the Canadian Armed Forces:

(a) The surveillance of our own territory and coast lines, i.e., the protection of our sovereignty;
(b) The defence of North America in cooperation with US forces;
(c) The fulfilment of such NATO commitments as may be agreed upon; and
(d) The performance of such international peacekeeping roles as we may from time to time assume.

Maritime forces have roles to play in all four areas. To begin with the roles which are clearly military but nevertheless necessary in peacetime, there is the surveillance of waters (including ice-covered waters) for submarines or other foreign military activity, and the contribution to NATO's Standing Naval Force Atlantic. In the twilight between peace and war there is support of United Nations peacekeeping, maritime support of NATO flexible response, and the delivery and supply of an air/sea transportable force to the NATO Northern Flank. In the event of war, it would be necessary to conduct surveillance and control of the waters in the vicinity of Canada for missile submarines, attack submarines, warships, aircraft, and hostile activities by trawlers. It would be necessary to provide protection for friendly shipping, including mine countermeasures, and to escort task forces.

In peacetime there are a number of necessary non-military maritime activities to be conducted by the Canadian government, many of which are suitable for the military forces acting in concert with other departments. In this paper, the term "Canadian Maritime Strategy" will be interpreted to include consideration of national maritime activities that may be essentially non-military. It is a considerable list, on which we find seaborne trade, fisheries, navigation, Arctic resupply, icebreaking, ice reconnaissance, provision of weather information, search and rescue, control of pollution, control of exploitation of the seabed, control of customs and immigration, cable repair, and oceanographic research.

The principal non-military activities will now be discussed before returning to the military activities.

Non-Military Maritime Activities

Seaborne Trade and Commerce

About half of all the goods produced in Canada are exported. About 35% of Canadian exports and 29% of imports are with countries overseas, nearly all being carried in ships. It is evident that a truly vital interest of Canada is that this trade be able to continue in a safe and efficient manner.

We have been told that in the 1920s Admiral Sir Herbert Richmond considered the basic requirement for fighting ships to be the protection of the merchant fleet. In the 1970s it seems to be clearly in the interests of all the important powers that merchant shipping should operate unmolested. It seems probable that common interest will ensure that commerce flows at sea unless international disagreements reach a very dangerous state indeed. And those areas where interference with merchant shipping by armed force might occur in situations short of global war are so far from Canada that our own maritime strategy does not need to place much emphasis on protection of commerce except in the event of a major crisis involving our allies as well as ourselves.

Food from the Sea

The ever-increasing population of the world produces a corresponding need for more food supply. Fish is a particularly desirable food because of its high protein content, an essential component of a healthy diet not easily or cheaply supplied through agriculture. Modern methods, including scientific search for fish and the provision of large factory ships moving with the fishing fleet, enable enormous catches to be taken. Since 1938 the world fish catch has more than tripled. But the resources of the ocean are not limitless, and the continued harvest of fish of several important species is already endangered. It is evident that the overall well-being of mankind would be improved by controlling the locations, types and quantities of fishing in such a way as to limit the catches to match the "maximum sustainable yield." Efforts to arrange this by international agreement failed to save whales and the whaling industry, but progress is being made in fishing. It may be that the problem can be solved by international agreement. However, since some of the best fishing grounds in the world are close to Canada, though beyond territorial waters, our maritime strategy must take into account the need to protect the interests of our fishermen, who rank in the first three among world exporters. Support could take the form of action against fishermen of another country not recognizing rules established by Canada, or of joint action by an international force to enforce rules agreed by their members but disobeyed by individuals or by fishermen of non-signatory countries.

Prevention and Control of Pollution

Oil pollution at sea is a cause of concern for Canada, especially with the discovery of oil in the Arctic, the dangers of Arctic navigation, and the delicate Arctic ecology. However, in addition to the problem of heavy pollution following an accident to a tanker or leakage from a submarine oilwell, there is also a need to prevent the careless or intentional deposit of oil, garbage, or other pollutants from ships in coastal waters. Accidents may be prevented by insistence on adequate standards of construction and navigation, intentional transgressions by the expectation of identification and legal action. After an accident, prompt measures by properly equipped teams may prevent or greatly reduce pollution, or expedite the cleanup. All of these are maritime responsibilities, though not primarily matters for the Department of National Defence, and they have been increased in magnitude by the recent passing of the Arctic Waters Pollution Prevention Act, an initiative not approved by several of the world's major shipping nations.

Control and Regulation of the Exploitation of Offshore Mineral Resources

It is becoming increasingly evident that the valuable mineral deposits existing on and under the surface of the earth are distributed on continental shelves as well as on dry land. The sequence of prospecting, drilling, mining and removing of minerals may be more difficult and expensive on the seabed than on land, but it will be carried out with great economic profit in the coming years. Since there is a greater area of continental shelf adjacent to Canada than to any other country except the Soviet Union, we have a tremendous stake in the matter.

At present, activities appear to be proceeding in accordance with Canadian law, with prospectors and drillers applying for government licences. Their financial investments are so great that it would appear in their interests to obey all regulations meticulously as long as the costs are reasonable. However, questions may arise regarding jurisdiction in areas not clearly on the continental shelf, or disputed by two or more nations. If military installations are ever built on the seabed (such as depots for submarines) there could be an interaction between defence and civil activities, even to the extent of arms control inspections being demanded by international bodies.

Canadian maritime strategy must take account of the economic importance of the seabed, especially on our large continental shelf, and of the likelihood that international disputes are going to arise concerning jurisdiction on the ocean floor.

Other Non-Military Maritime Activities

Search and rescue, both on sea and land, occupies a considerable effort in flying time of aircraft and, on occasion, in diversion of ships. Operations are coordinated by the Department of National Defence and the Ministry of Transport. Some aircraft are specially equipped for this role.

Several government agencies are involved in the safety of shipping, for which it is necessary to provide navigation aids, charts, meteorological information, wharf maintenance, dredging, and many other services. Icebreaking, Arctic resupply, and ice reconnaissance are other important services which will probably need to be expanded considerably as activity in the Arctic increases.

The regulation of customs and immigration is mainly a matter of enforcement on land, but is supplemented by a fleet of small vessels operated by the Marine Division of the RCMP.

Research and data collection for hydrographic, oceanographic, fisheries, and defence purposes is carried out by several departments. To date, oceanographic research cruises have not been subject to international restrictions, but there are moves on the part of some of the underdeveloped countries to control the extent of the surveys which can only be done by nations possessing advanced equipment.

The Law of the Sea

The law of the sea has developed over a long period in which the free movement of seaborne commerce was desired by nearly all important countries, and in which there were enough fish for everyone, although it might be necessary to go far from home to find them.

In the future it seems probable that a nearly universal wish for safe and easy passage of merchant shipping will continue. However, as has already been indicated, a number of new factors are emerging which are likely to cause serious conflicts of interest among nations. It may not prove possible to obtain international agreement to modified laws, and disputes are likely to arise regarding boundaries of jurisdiction. Of concern to Canada is the status of the passages between the islands of the Arctic Archipelago, the boundary of jurisdiction on the seabed between Newfoundland and St. Pierre and Miquelon, and between Nova Scotia and Maine. Legal jurisdiction over ice floating on the sea is not certain.

It is not suggested that Canadian claims will be established by winning naval battles. But it is suggested that when laws are in dispute it may be necessary to conduct surveillance and inspection in order to be aware of activities and to uphold claims by national presence with a capability for defence and enforcement.

Military Maritime Activities

The Support of Strategic Nuclear Deterrence

There can be little doubt that the central theme of Western military strategy, and very probably also that of Eastern, is maintenance of stable strategic nuclear deterrence. The balance of deterrence depends on three offensive systems (bomber aircraft, Intercontinental Ballistic Missiles, and missile-firing submarines) and four defensive systems (air defences, ballistic missile defences, antisubmarine defence, and civil defence). Air defences can use airborne early warning systems flying over the sea, and ballistic missile interception systems of the future may be based on ships or aircraft flying over the sea. However the two systems of purely maritime character are the missile-firing submarines and the defence against the missile-firing submarines.

The most effective missile-firing submarines are nuclear-powered and carry sixteen ballistic missiles (submarine-launched Ballistic Missiles, or SLBMs) which can be launched underwater. These nuclear-powered ballistic missile submarines are designated as SSBNs. They usually operate alone, and if some sort of protective escort were desired it would probably take the form of nuclear-powered attack submarines (SSNs). Canada does not contemplate any role in the operation or escort of SSBNs.

An interesting debate can be produced regarding the contribution of antisubmarine defence to the preservation of stable nuclear deterrence. It starts with the hypothesis that a state of mutual deterrence exists if both opponents possess a force of offensive weapons sufficiently numerous, invulnerable and effective that no matter what attack (the first strike) may be made on them, enough will survive to be able to retaliate (the second strike) against the attacker's cities and industry to a degree beyond what could be endured. The mutual deterrence is also said to be stable if its existence would not be jeopardized by small changes in the forces (or the effectiveness of the forces) on either side, if neither side is required to launch on warning (i.e., before an attack has actually been delivered), and if neither side has any rational motive to attack first (i.e., a pre-emptive attack to prevent some action by the adversary). In general, steps to increase the certainty of retaliation in a second strike are stabilizing, while steps which might make it possible for a counterforce first strike to disarm the opponent and make it impossible for him to retaliate are destabilizing.

The case against antisubmarine defence is based on the fact that SSBNs are such an effective weapon system for second-strike counter-value retaliation. Once at sea they are invulnerable to the opponent's strategic offensive weapons. Since they are smaller and less accurate than ICBMs, SLBMs are less suitable for a counterforce strike against hardened point targets such as ICBMs in their silos, but they are quite large and accurate enough to wreak unbearable damage on urban or industrial targets. Therefore, since SSBNs are better for retaliation than for a counterforce first strike, they are stabilizing, and measures to oppose them are destabilizing.

The case can be elaborated by three additional arguments. First, there are more ICBMs than SLBMs, and there is not much defence against ICBMs. There is little value in trying to defend against a minor threat until something effective is available against the major threat. Second, submarines are difficult to locate and track, especially when they proceed slowly and silently. Third, even if defences succeeded in locating and tracking SSBNs, they cannot attack them in peacetime in international waters. And if a surprise first strike were launched at a predetermined instant, all the missiles would be gone in a few minutes, after which an attack on the submarine would be too late.

The opposing argument is based on the high vulnerability of certain elements of the retaliatory system to surprise attack, the use they can make of early warning to reduce this vulnerability, and the fact that the trajectory of an SLBM is much shorter and lower than that of an ICBM, particularly if the submarine comes close to shore before launching. The elements in question are bomber aircraft, especially those based near the coast, and command and control centres. An SLBM burst above an airbase would destroy all the bombers on the ground. However, strategic warning (of days or hours) would permit dispersal of aircraft to many bases, mainly inland, while tactical warning (of a few minutes) would allow some of the aircraft to save themselves by taking off before the missile exploded. Command and control centres can also take steps to ensure the continuation of their function by such measures as dispersing key personnel to alternate posts (some may be airborne) or into protected locations. Then, if antisubmarine surveillance gives warning of a buildup in SSBNs off the coast, or of movements towards the coast, steps can be taken to reduce vulnerability and ensure the capability to retaliate. Even prompt notice of missile launching can result in saving of retaliatory capability. So, of course, would the interception of SLBMs in flight or the destruction of SSBNs before they had launched all of their missiles.

In answer to the other arguments, the proponents of antisubmarine defence point out that the current rate of building of Y-class SSBNs by the Soviet Union will take them well past the US total (656 SLBMs) by the mid-1970s, while ballistic missile defence will begin to oppose the unchallenged freedom of the ICBMs. And, while location and tracking of submarines is difficult, it is not impossible, and is likely to improve with research and experience.

A study of the geography of the North Atlantic shows that most of the firing positions close to bomber bases are closer to the USA than to Canada, but that many of the likely transit routes to these positions come through waters closer to Canada than to any other country. It could well be that a sensible role on which Canadian maritime forces could concentrate would be surveillance and tracking of SSBNs transiting through these waters close to Canada. This function is clearly a stabilizing one, and one more easily done from Canadian bases than others. If effective surveillance in the deep ocean drove SSBNs to choose circuitous transit routes through shallow coastal waters, there would be a case for antisubmarine surveillance on the Pacific and Arctic as well as the Atlantic coasts. Surveillance under the Arctic ice could require the development of new techniques which could be useful for civilian as well as military applications.

Protection of Shipping: Supply and Resupply on the North Atlantic

The great question to be asked in this connection is how likely is it that hostilities could remain at the very high level at which shipping on the North Atlantic was subject to non-nuclear attack, whether by submarines, aircraft, or surface ships, for a long period, without the situation escalating to full nuclear war? Once all-out nuclear war breaks out, it seems most improbable that hostilities will continue for a long period thereafter, with NATO's fate depending on the maintenance of the Atlantic lifeline.

Another non-nuclear Battle of the Atlantic would require a large force of escort vessels equipped for antisubmarine and antiair warfare, and another force of mine countermeasures ships. Canadian ports and airfields would be very important. But it does not appear to be a very probable eventuality.

Limited Nuclear War at Sea

The suggestion has been made that the most likely place for a war to escalate to the level at which tactical nuclear weapons are used in combat between military forces, but to remain at that level, is at sea. Here the line of demarcation between tactical and strategic use is fairly clear, the chance of inadvertent wholesale destruction of large communities is minimal, and the number of civilians endangered would not be large. The effectiveness of antisubmarine warfare would be significantly increased by the employment of nuclear weapons.

Support for NATO's Flexible Response

NATO's plan is to rely on strategic deterrence to prevent general nuclear war, and to be able to produce a flexible response adequate to any provocation short of general nuclear war. Maritime forces are well suited to flexible response. However, it is likely to be in the European theatre that the response would be made, and the present Canadian maritime forces are not well suited to work in a hostile air environment or with fast carrier strike forces. We do, however, contribute a destroyer to SACLANT's Standing Naval Force Atlantic. We are committed to send by air a battalion group to Allied Command Europe's Mobile Force Land if the latter is deployed to Denmark or Norway, and to send the balance of an air/sea transportable combat group from Canada to the northern flank in the event of an emergency. With our present maritime forces we could supply antisubmarine escort and air surveillance.

United Nations Operations

The United Nations operations in the past have been primarily on land. It is not, however, impossible to imagine an operation against an island or other area largely dependent on supply by sea. Economic sanctions could take the form of a partial or complete maritime blockade. There is a precedent for this in the UN trade sanctions against Rhodesia, in which the Royal Navy has attempted to intercept tankers carrying oil for shipment through Beira to Rhodesia.

The Present Strength of the Canadian Military Maritime Forces

Professor Schurman mentioned the importance of materièl in the development of naval strategy, and warned of the danger that, in the absence of a strategic doctrine, decisions will be dominated by questions of materièl. This is especially true in a small navy in times of austere budgets. Remembering the long lifetime of maritime equipment, it is inevitable that today's forces are the result of a past strategy, and that it will be a long time before a change in strategy today can be reflected in forces with radically different equipment. The saving grace is that maritime forces are inherently versatile and flexible.

For those not familiar with them, it may be worth a few minutes to sketch the structure of the present Canadian military maritime forces.

There are twenty destroyers, displacing about 2900 tons, rather slow, and primarily equipped for antisubmarine warfare. About half of them carry one Sea King antisubmarine helicopter, with a crew of four, and equipped with dipping sonar and torpedoes. All have the Limbo mortar for antisubmarine depth bombs, most carry antisubmarine torpedoes, and some have Asroc rocket launchers for antisubmarine torpedoes. Most have variable depth sonar. All have 3-inch automatic anti-aircraft/surface guns.

Four new DDH-280 class destroyers will join the fleet in 1972/73. These are 4000 ton vessels, each carrying two Sea King helicopters. They also have Limbo, AS torpedoes, a 5-inch gun, and Sea Sparrow close range surface-to-air missiles.

We have three Oberon-class diesel-powered attack submarines and one old Tench-class submarine. Two 22,000 ton Operational Support Ships carry three Sea Kings, two 3-inch AA guns, and Sea Sparrow. One 23,000 ton helicopter and supply ship carries six Sea Kings.

A very important component of the Canadian maritime forces are the 32 Argus long-range patrol aircraft. With a 15-man crew and very long endurance, these carry a large radar, sonobuoys, antisubmarine torpedoes, magnetic anomaly detectors, and other equipment for maritime reconnaissance. There are in addition shorter-range Grumman Trackers, initially procured for carrier use. They have a crew of four, and carry sonobuoys, torpedoes and rockets.

Possible Future Capabilities

In its 1970 report respecting maritime forces[5], the Commons Standing Committee on External Affairs and National Defence recommended the following capabilities for Canadian maritime forces in the period 1973-83:

1. considerable surface and subsurface surveillance and identification capability,

2. limited surface and subsurface tracking and localizing capability,
3. limited surface and subsurface challenge and destruct capability,
4. limited self-defence capability.

In respect of new equipment, they recommended:

1. the continued maintenance of long range airborne maritime patrol forces to provide considerable surveillance and identification as well as limited localizing, tracking and challenge and/or destruct capabilities;
2. the maintenance of surface forces, with the emphasis on light and fast general purpose vessels to provide limited surveillance as well as limited localizing, tracking, and challenge and/or destruct capabilities.
3. careful consideration of the possibility of developing and deploying in appropriate locations in Arctic regions bottom-based systems providing these are found to be capable of effective surveillance and identification under ice;
4. no acquisition of nuclear-powered submarines, given the high estimated cost.

Over the long term, our maritime strategy will depend on our answers to several major questions:

1. What part will Canada elect to play in opposing the missile-firing submarine? (Surveillance? Attack the submarine? Intercept the SLBMs?)
2. What will be the requirements for the maritime support of NATO?
   (a) Is the transatlantic convoy, opposed by submarines, aircraft, and surface ships still an important possibility?
   (b) Will there be a requirement to mount and protect smaller task forces?
   (c) Will Soviet expansion into new areas, or perhaps the increasing dependence of developed Western countries on petroleum imports create new naval tasks?
3. What will be the requirements for international missions such as UN peacekeeping, protection of nationals in time of insurrection, aid to small Commonwealth countries requesting assistance to restore order, etc?
4. What will be the Canadian domestic requirements?

Another, more technical question is "should we continue to design a maritime force with special capabilities in the antisubmarine role, or should we now aim at a more versatile general purpose force?"

These are many questions indeed. But it would take a real optimist to predict that a country with immensely long coastlines on three of the world's great oceans will be able to maintain all of its rights and interests in the turbulent seventies (likely to see many clashes of interest on and under the sea) solely by the efforts of diplomats, lawyers and disarmers, with no requirement at all for some type of sea-going policeman. Indeed, the work of the diplomats, lawyers and disarmers is likely to be aided by the right type of maritime forces, including those of the smaller as well as the larger countries. The task before Canada's maritime strategists is to identify the right type of force, and to persuade our authorities to create it in time.

7 Problems of Naval Arms Control

The High Seas, The Deep Ocean
and the Seabed

WILLIAM EPSTEIN
Disarmament Affairs Division of the United Nations

Diversified Approach

In addition to the traditional responsibilities which maritime forces have discharged for centuries, naval forces today, particularly those of the two super powers, are rapidly taking on new and increased responsibilities. First and foremost amongst these in the nuclear age is the function of preserving the global nuclear strategic balance. The very importance of this function has led to an expansion in detection and surveillance activities. Navies are also beginning to become involved in oceanographic research for scientific and commercial purposes. In the future one may expect a broad spectrum of responsibilities in connexion with various civil-military activities in the protection of commercial and scientific activities on the seabed and the ocean bottom, the undertaking of rescue operations and, hopefully, the verification of various naval arms control measures and agreements.

Because of the diversity of today's naval activities, the question of naval arms control cannot be viewed as a single unified problem. The consideration of naval matters generally, as well as of the arms control aspects, has been divided along functional and geographic lines. Questions affecting the global strategic balance, involving nuclear missile submarines, are being discussed at SALT (the Strategic Arms Limitation Talks) between the United States and the Soviet Union in Helsinki and Vienna. The more traditional questions of naval arms control on the high seas are being discussed in various forums: those concerning regional naval arms control both nuclear and conventional, for the Mediterranean, the Indian Ocean, the Caribbean Sea and the oceans surrounding Latin America, have been discussed in regional groupings and also in bilateral talks between the Soviet Union and the U.S.; they are of course also discussed in the United Nations in the context of both disarmament and the law of the sea. A new (third) U.N. conference on the law of the sea is scheduled to be held in 1973; many questions such as the breadth of the territorial sea, of sovereignty over the continental shelf, of rights of passage through straits, have become of increasing importance. Questions relating to the seabed and the ocean floor are being discussed in the U.N. – in the U.N. Committee on the Peaceful Uses of the Sea-Bed and the Ocean Floor beyond the Limits of National Jurisdiction, and also by the General Assembly which has on its agenda an item with the unwieldy title of "Reservation exclusively for peaceful purposes of the seabed and the ocean floor and the subsoil thereof underlying the high seas beyond the limits of present national jurisdiction and use of their resources in the interest of mankind, and convening of a conference on the law of the sea." The question of denuclearizing or demilitarizing the seabed and the ocean floor has been and is the subject of discussion in both the Geneva Disarmament Conference and in the United Nations.

The fragmentation of the problem of naval arms control into a multitude of subjects and their consideration in a wide variety of forums, has made it difficult to obtain any general overall view of this complex field. There are, however, some discernible common strands which can be pulled together.

Global Strategic Balance

The newest and most important area of the exercise of naval power is that of the role of nuclear submarines in the maintenance of the global strategic balance.

While the two nuclear super-powers, the U.S. and the Soviet Union, still seem to base their defence and deterrence postures on the "triad" of strategic nuclear weapon systems – bombers, land-based missiles and submarine-based missiles – it seems clear that the strategic bombers and land-based missiles are becoming obsolescent. Their survivability in case of attack is becoming increasingly doubtful and strategic thinking is turning more and more to submarine systems. The submarine nuclear forces, unlike bombers and land-based missiles, appear to be almost completely invulnerable to any first nuclear strike or counter-force attack, thus ensuring the reliability of the nuclear deterrent. Hence military analysts are increasingly calling for primary, if not exclusive, reliance on submarine-based nuclear systems to preserve the nuclear deterrent.

So far as is publicly known, the present generation of submarines can submerge to a depth of around 1,000 feet and the furthest range of their missiles is some 3,000 miles. Long-range plans are already on the drawing boards for what is known as ULMS (under sea long-range missile systems) which envisage submarines which can go down to a depth of 10,000 feet and whose missiles would have a range which could hit a target anywhere on earth.

At the present time, the U.S. has 41 Polaris-type submarines, each of which is capable of firing 16 nuclear missiles, making a total of 656 missiles. Thirty-one of these submarines are being converted to Poseidon missiles of the MIRV (multiple independently-targeted re-entry vehicles) type, reported to carry 10 to 14 nuclear warheads, which would thus provide some 5,000 sub-based nuclear warheads. At the Moscow summit meeting it was agreed that the USSR would limit its nuclear missile launching submarines (known as Y-class submarines) to 45, also carrying up to 16 missiles each, with a total of 710 submarine missiles. (Western sources claim that these missiles have a range of only some 1,750 miles and they are not MIRVed, but are single warhead missiles, but it is clear that the Soviet Union will in a few years catch up to the United States in this respect.)

By a Protocol, dealing with nuclear submarines and submarine launched ballistic missiles (SLBM's), to the 5-year Interim Agreement on Offensive Nuclear Weapons signed at Moscow as part of the SALT agreements, it was agreed that the U.S. could have a total of 44 nuclear missile submarines with no more than 710 SLBM's, and the USSR could have a total of 62 nuclear missile submarines with no more than 950 SLBM's. The increase in the number of submarine missiles was to be matched by an equivalent decrease in land-based missiles.

The SALT agreements signed at Moscow on 26 May 1972 fixed numerical limits or quantitative ceilings on both defensive and offensive strategic nuclear weapons, but imposed no qualitative restrictions. There is, therefore, nothing at this stage to prevent either side from replacing its nuclear submarines and SLBM's with more advanced models. The freeze or ceiling on numbers is not accompanied by any freeze or ceiling on technological and scientific improvements.

In January 1972, the U.S.A. Administration requested a sum of almost $1 billion for preliminary work on ULMS. Some have agreed that this would provide the U.S. with a useful "bargaining chip" or "hedge" in its further negotiations with the USSR. Others have argued that the U.S. should in any case go ahead to produce the new generation of ULMS submarines to guarantee the defence of the U.S. and the preservation of the nuclear deterrent. Those favouring this approach hold that the ULMS would be even less vulnerable than Polaris submarines and that, even if the Soviet Union also proceeded with the production of a new generation of nuclear submarines, this would tend to stabilize the balance of mutual deterrence. Others reply that this would merely trigger a new, expensive and unnecessary round in the fantastic nuclear arms race, that the present Polaris and Poseidon systems and the Soviet submarine systems are more than enough to ensure the stability of the nuclear deterrent and that it would be a form of sophisticated insanity to embark on another spiral in an unending and futile nuclear arms race.

In any case, Secretary of Defense Laird in June 1972, after the conclusion of the Moscow SALT agreements, proposed that the U.S. go ahead with the building of 10 new Trident submarines, a new generation submarine costing about $1 billion each, more advanced than Polaris or Poseidon. He also asked for a new "cruise" or guided missile for launching from submarines. It is not easy to understand the purpose of these new nuclear weapon systems since they would not affect the invulnerability of the other side's deterrent.

At the present time, it would seem that advantage could be taken of the numerical limitations agreed at Moscow in order to work towards a qualitative freeze as well on nuclear weapons. Such a freeze, together with a limitation on bombers and land-based missiles, would put an end to both the quantitative and qualitative nuclear arms races, while ensuring the continuance of the present balance of nuclear deterrence.

Anti-Submarine Warfare (ASW)

Some people have questioned whether the present submarine nuclear strategic forces are really so invulnerable to ASW attack, and whether on the "worst case hypothesis" it might not be safer to develop and deploy newer and more sophisticated generations of submarines. It seems very clear, however, that there is no effective ASW capability at the present time and that prospects for the achievement of such a capability are extremely remote. Such a capability would require the detection of every enemy missile-carrying submarine in all the oceans, their positive identification (discriminating between enemy and friendly subs, schools of fish and other false targets), fixing their precise location and, finally, launching a simultaneous attack which would destroy practically the entire enemy force. If even a few submarines could escape such destruction, they would retain sufficient nuclear retaliatory capacity for a second strike which could largely destroy the territory of the attacking state. No nation can even begin to approach such a capability.

Under-sea detection is based on sonars of two types — "passive" sonars (which listen to and record the sound created by some other submarine's propulsion machinery or its movement through the water), and "active" sonars (which send out high energy waves which strike underwater objects and return echoes to the listening instruments). The range of both types of sonars is limited at the present time to a few miles. Passive sonars may fail to detect sufficiently quiet and slow moving submarines. Active sonars can seek out targets and also determine distance better than passive sonars, but they give themselves away. The sound waves generated by the transmitter reveal its presence over a far greater area than its own area of detection. Thus, a nuclear submarine with passive sonars can detect an adversary's active sonar before it is itself detected and thus have time to get away or strike first at the opponent. Any detection system would have to include both active and passive sonars. These would require a grid of high-power transmitters and detectors covering the entire area under surveillance. It would be impossible to build an oceanwide ASW system without it being very visible to the other side. The very size and extent of the system would make feasible the banning of such systems, since verification would not be difficult; an oceanwide sonar system could not be built surreptitiously. Moreover, even if such a network were built, counter measures would be possible such as jamming the network and creating other unmanageable problems of discrimination.

Thus the invulnerability of submarine systems can be assured in the future by an agreement to prohibit the deployment of high-powered transmitters for active sonars. Whether or not the American or Soviet Governments might eventually agree to a ban on such high-powered transmitters is far from clear. Both sides seem more interested at the present time in continuing their ASW research and development than in limiting their efforts in this field. It has been estimated that the U.S. is spending approximately $2 billion annually for ASW, which is roughly comparable to what it spends for under-sea weapons. This would seem to indicate its continuing interest in this field.

Even if one side should make some significant break-through in the field of ASW, for the reasons outlined above it could hardly hope to eliminate the entire submarine force of the other side in a first strike. Thus it would always be subject to the assured destruction capability of a retaliatory second strike by the surviving submarine forces.

Geographical Limitations

Two treaties have been recently concluded which place geographical or regional limitations on the deployment of nuclear weapons in the marine environment.

The Treaty for the Prohibition of Nuclear Weapons in Latin America (the Treaty of Tlatelolco) of 1967, which provides for the creation of a nuclear-free zone in Latin America and the Caribbean, includes a vast area of the Atlantic and Pacific Oceans in the zone. The area of this zone is similar to that declared as the naval security zone for Latin America during World War II. The U.S. and the U.K. have signed and ratified the Protocol to the Treaty thus pledging themselves to respect the nuclear-free status of the zone. France and China have declared their sympathetic attitude towards the nuclear-free zone, but the USSR has indicated that it does not intend to sign the Protocol. The U.N. General Assembly has adopted several resolutions calling on all the nuclear weapon powers to sign and ratify the Protocol.

The Treaty on the Prohibition of Emplacement of Nuclear Weapons and other Weapons of Mass Destruction on the Seabed and the Ocean Floor and in the Subsoil thereof (the Seabed Treaty) was signed in February 1971. This

Treaty provides that no nuclear or any other weapons of mass destruction would be placed on the seabed and ocean floor beyond a 12-mile coastal zone. Further details about the negotiation and conclusion of this Treaty are given below.

Last year Ceylon, together with a number of non-aligned countries, proposed that the Indian Ocean be declared to be a "peace zone." Not only was it to be made a denuclearized zone but the proposal was intended to demilitarize the entire Indian Ocean. After much discussion, in which questions were raised concerning the freedom of the high seas and existing rules of international law, the General Assembly adopted a resolution calling upon the great powers to enter into consultations with the littoral states to halt the expansion of big power presence in the region and to eliminate all bases, military installations, weapons of mass destruction and any other manifestations of great power military rivalry from the Indian Ocean. The proposal also called upon the major maritime users of the ocean, as well as the littoral and hinterland states, to consult in order to ensure that warships and military aircraft do not use the Indian Ocean for the threat or use of force against the territory of states in the region. The resolution was adopted by a vote of 61 to none, with 55 abstentions, which indicates the doubts entertained by a large number of countries. The U.S., U.K., France and the Soviet Union all abstained on the vote, but China and Japan supported the resolution, as did most of the non-aligned states.

The question will be the subject of much debate at future sessions of the General Assembly.

The Seabed

The question of preserving the seabed and ocean floor and the subsoil thereof for peaceful purposes first arose in 1967, because of the growing scientific and commercial interest in those areas and because of fears that the nuclear powers might wish to place nuclear weapons there. The United Nations called for the use of the seabed and ocean floor exclusively for peaceful purposes and set up a Committee on the Peaceful Uses of the Sea-Bed. This Committee has been working to establish an international régime for this new environment.

In the negotiations for the Seabed Treaty, mentioned above, at the Geneva Disarmament Conference the Soviet Union called for the complete demilitarization of the seabed outside of a 12-mile coastal zone, with verification and inspection by the adversary parties. It claimed that it would be easy to verify a complete demilitarization as every installation on the seabed could be open to inspection to see that it was not for military purposes. The U.S., on the other hand, called merely for the denuclearization of the seabed, claiming that it would be too expensive to set up conventional arms on the seabed, that the latter weapons could not threaten foreign territories, that detection and surveillance devices were essentially defensive and not offensive, and that the difficulties of verifying a total demilitarization would be great because it would involve discriminating between what were peaceful or military installations. The U.S. also wished the coastal zone to be limited to three miles, which is the width of the U.S. territorial sea.

Canada put forward a compromise suggesting a 200 mile security zone where defensive activities would be permitted by the coastal state, and all offensive military activities would be prohibited for the entire seabed area beyond a 12-mile coastal zone. However, the U.S. and USSR did not accept the Canadian proposal which would give the coastal states special privileges in a 200-mile security zone. Canada and other smaller maritime and coastal states also urged that interested states should be able to participate in the verification carried out by the great powers and should have the right to call on the great powers directly or through the United Nations to assist them in verification procedures.

Finally, a Treaty was agreed upon calling for the banning of nuclear and all other mass destruction weapons on the seabed beyond a 12-mile zone. The Treaty also contained provisions safeguarding the rights of coastal states and procedures for international verification.

From the strictly military point of view, the seabed, the ocean floor and the subsoil thereof, have less value than does the ocean itself. Mobile systems, such as submarines, have a much wider range and manoeuvrability and probably would cost a good deal less. From the political point of view, it is interesting to note that the coastal states and the smaller powers asserted their interest and the great powers found it useful to compromise and make concessions. They may have been helped in this respect by the argument of some of the smaller powers and coastal states that, unless the Treaty took full account of their interests, it might be left solely to the big powers to agree without the participation of the little powers who, in any event, had little likelihood of being able to place nuclear or other weapons of mass destruction on the seabed.

As regards the future, it is quite possible that there will be some continuing rivalry of a commercial nature, rather than of a military one, for control of the resources of the seabed. Hence the need for an international régime which is now being sought by the United Nations. It is not easy to visualize early agreement on further measures of demilitarization of the seabed because of the political complexities. It is difficult to specify measures that might be acceptable to both the great and small powers. From the military point of view, this may not be too urgent a problem because, as many experts point out, it is hardly worth the trouble of establishing military installations on the seabed, since they would probably be less effective and more expensive than mobile systems in the oceans or on the high seas.

Conventional Naval Forces

There are different views as to whether a naval arms race exists, or is likely to occur, in what are known as "conventional" naval forces (to distinguish them from nuclear forces). There are also different views as to the importance or otherwise of such a naval arms race.

There is fairly wide agreement that naval vessels are of limited *military* use against each other. *At sea,* the navies are primarily symbolic and political in their purpose. They have mainly a deterrent effect — to persuade the other side that a planned course of action may be too risky. The regional deployment of ships may be much more important than the number of vessels or guns. By being on the spot or by being able to get first to a region of turmoil or crisis, a naval force can confront an opposing force with the difficult choice of staying away or risking involvement in possible hostilities. Naval forces do, however, have some military as well as political effectiveness in supporting operations *on land*; a naval power can intervene in local or regional conflicts by supplying materials and equipment, by moving men, by bombing and shelling, or merely by threatening to use force.

Because of these capabilities, it is possible that a conventional naval arms race may occur, despite, or perhaps as a result of, the nuclear naval arms race being largely stabilized on the basis of mutual deterrence. At the present time there appears to be greater competition in the deployment of navies in different parts of the oceans — the Mediterranean, the Indian Ocean and the Far East — than in increasing the numbers and the over-all size of naval forces. In fact, for a number of years there has been a decrease in the number of naval vessels.

There is, however, a continuing naval arms race in qualitative if not quantitative terms. There is also a considerable debate in the United States and other Western countries as to whether the increasing presence and expansion of Soviet naval forces in different parts of the world is or is not a serious threat to Western political as well as military and commercial interests. Hence a naval arms race among the great powers, and particularly the USSR and U.S., in conventional forces cannot be ruled out.

During the 1920s and 1930s, efforts to control naval forces, all of which were of course non-nuclear, were based upon formal global limitations — by numbers, by tonnage, by categories of vessels, or by gun calibers. In the nuclear age, except for nuclear submarines, the overall size and categories of naval vessels have become less important than questions of geographical deployment. Current efforts towards naval arms control in the conventional field have been concerned more with the limitation or regulation of the unrestricted freedom of activities on the high seas and the ocean environment. The present tendency appears to be away from general and universal treaties and towards bilateral agreements, tacit understandings and the exercise of mutual and unilateral restraint.

The Soviet Union and the U.S., for example, reached some naval "understandings" in October 1971. They reaffirmed their recognition of the freedom of the high seas and the right of surveillance of each other. They also agreed to take measures to ensure the safety of their naval forces in near proximity. This agreement was confirmed by the signing on 25 May 1972, at the Moscow Summit Meeting (and entry into force on the same day) of an agreement to prevent incidents at sea and in the airspace over it between vessels and aircraft of the U.S. and Soviet navies. The agreement not only laid down rules and procedures for ships and aircraft operating in close proximity, but also provided for the creation of a committee to consider additional specific measures and for annual consultations to review the implementation of the agreement.

In June 1971, Mr. Brezhnev stated that the Soviet Union was prepared to solve the problem of competition with the U.S. in the Mediterranean Sea and Indian Ocean. In January 1972, Washington announced that it had approached the Soviet Union concerning the possibilities of mutual restraint in the development of naval forces in the Indian Ocean. In the Mediterranean, however, at least up to the present, rivalry between the Soviet Union on the one side and the U.S. and NATO powers on the other, appears to be continuing. Each side appears to be seeking to strengthen its position. But, even here, it has been suggested that the rivalry or competition is related less to purely military objectives than to political ones linked to the Arab-Israeli problem.

At the present time there are no specific efforts or discussions to achieve general or global naval arms control. The problem is being approached either in a piece-meal fashion in bilateral or regional discussions or has been relegated for future consideration as one of the items in a programme of general and complete disarmament.

Unilateral acts of restraint, bilateral understandings and tacit agreements do not of course have the force or validity of binding international or multilateral agreements, but they are very useful instruments for avoiding the risks of miscalculation and accident that are always possible in situations of rivalry and potential conflict.

It is likely that the long-standing suspicions and competing interests and rivalries among the great maritime powers will continue despite the recent efforts to keep them from getting out of hand. Whether or not the naval competition will increase in the future depends on so many unknown political variables as to make speculation not only hazardous but perhaps useless. In any case, however, the political effects of the SALT agreements and the freeze on nuclear submarines and missiles, with the resultant stabilization of the balance of nuclear deterrence, would greatly reduce the chances of any conventional war between the super-powers. While the rivalry in conventional naval forces may well continue and possibly even increase in the future, the possibility of actual naval combat on the high seas between the great powers appears to be becoming increasingly remote. Not only does the SALT agreement on submarines tend to diminish tensions, but it would seem to underline the need to prevent any confrontation from developing into large scale conventional naval hostilities. The agreement to reduce and prevent incidents between American and Soviet naval aircraft and ships would seem to reinforce this tendency towards restraint. Thus, while the prospects for any global or general conventional naval arms control agreements do not seem to be very likely, the possibilities for reaching additional measures of restraint and of bilateral or regional understandings or agreements of a limited nature seem to be brighter.

The Smaller Maritime Powers

The smaller maritime powers and coastal states have traditionally been apprehensive of big power political and military rivalries on the high seas and have been concerned not to become involved in these rivalries. In recent years, as the possibilities and need for exploiting the resources in and beneath the ocean have become apparent, their concerns have expanded to include fears of competition from the big powers in exploiting the resources off their coasts, whether in the sea, on the continental shelf or on the seabed.

The super-powers have recognized the need to limit the degree and the area of their rivalries and have begun to take measures to avoid or reduce the risks of any confrontation with each other. The smaller powers, for their part, appear to wish to put some limitations on the unrestricted freedom of action – military, political and economic – of the big powers, and have begun to seek some clearer definition and legal recognition of their own rights and interests. A number of Latin American countries for example, have declared that their territorial seas extend for 200 miles and claim sovereignty over that entire area. Only China, among the great powers, has shown any support for this claim to a 200-mile territorial sea.

A new situation concerning the entire ocean environment may be emerging as a result of the growing suspicion and possible conflict of interests between the great maritime powers on the one hand, and the smaller coastal states on the other. The latter cannot of course compete militarily or economically with the great powers, but wish to assure for themselves the benefits of the freedom of the seas. Stated in another way, they want to place limits on the unrestricted freedom of the great powers so that they can obtain the maximum utilization for themselves of the resources of the seas and continental shelves off their coasts. Their desires are bound to have some influence on the future trends not only of the law of the sea concerning economic and commercial matters and questions of sovereignty, but also as regards the exercise of naval power. The effects of this recent development have already been felt in the completed negotiations for the Sea-Bed Treaty and in the current negotiations for an international régime for the peaceful uses of the seabed and for the holding of another conference on the law of the sea.

Conclusions

1. The dangers from the possible threat of nuclear arms are, of course, greater and the need for their control more important than those of conventional armaments. Fortunately, however, it appears that it might be easier to establish some limitation and control of nuclear arms than over the broad and diversified complex of conventional naval armaments. The dangers from the latter, despite the incentives to obtain some margin of superiority in them, are not as great as in the case of nuclear weapons. Hence freer competition and the absence of control appear to be less of a threat to the world community than in the case of nuclear weapons. The risks posed by nuclear weapons to both human survival and human welfare, as well as the existence of the current situation of a balance of mutual nuclear deterrence, make agreements for the control of these weapons both more necessary and more possible.

2. The freeze or ceiling established in the SALT agreements at Moscow on nuclear submarines and SLBM's gives ground for hopes that it may eventually be possible to place a freeze or limitation on the qualitative improvement of nuclear submarines and their missiles. The invulnerability of nuclear submarine systems would seem to stabilize the balance of mutual deterrence and thus to facilitate agreements not only to put a limitation on the numbers of such submarines but also on the development of new generations. On the other hand, however, it would seem that for a variety of reasons, including the desire to enhance their deterrent role, the nuclear super-powers might wish to maintain the freedom of their submarines to travel and operate in the entire ocean environment without regional or geographical limitations.

3. While it is most unlikely that any effective ASW systems can be developed in the foreseeable future, and although the limitation of such systems would appear to be relatively easy, there are no present indications of any movement for their limitation. Just as a severe limitation on ABM's removes much of the incentive for more and better offensive missiles and MIRV's, so too a ban on large scale development of ASW systems may be the best way of preventing the development of improved models of nuclear submarines and missiles.

4. It is difficult to visualize further measures for the demilitarization of the seabed in the foreseeable future, but since conventional armaments are of lesser importance in this environment, there would appear to be no great urgency in this regard.

5. While there are no concerted efforts to achieve any general or global arms control measures for conventional naval forces, there would seem to be some definite possibilities for additional bilateral and regional agreements or tacit understandings on measures of restraint in order to reduce the dangers of conventional naval competition and confrontation between the great powers. It is also possible that limited regional agreements may be feasible not only between the great powers but also among the states of specific regions.

6. Differences and disputes among the major maritime powers and between them and the smaller powers and coastal states seem likely to continue. They may even intensify over the passage of time. There is no reason, however, to believe that any of these differences or disputes should lead to serious conflict. Fortunately the various aspects of these problems are under active consideration in the United Nations.

7. While the manifold problems of naval arms control are being tackled only in piece-meal fashion and in a variety of forums, the important fact is that they *are* being tackled. The whole complex of problems — economic, political, legal and military — of establishing rules and controls for activities on and in the high seas, the deep oceans and the seabed is being considered in detail and in depth in bilateral and regional discussions and, above all, in the United Nations, where all nations can make their views and interests known. So long as the international community is determined to grapple with these problems, there is hope that over a period of time the nations of the world may succeed in establishing some effective international régimes for these environments so that they can be used for the benefit rather than the detriment of all nations and all men.

8 The Seas in the Seventies

LIEUTENANT-COMMANDER A. D. TAYLOR*
Department of External Affairs, Ottawa

The conduct of maritime strategy is taking place more and more in the foreign ministries, the legislatures and the international conference halls.

This paper reports on the current state of the Law of the Sea, particularly the legal issues as they have been and are being developed in the series of United Nations' Conferences on the Law of the Sea, and, using Canada as an example of a state that is both a *maritime* and a *coastal* state, illustrates how the interests of such a state are balanced.

The Law of the Sea Conferences

At the conclusion of the United Nations Conference on the Law of the Sea in Geneva on 29 April 1958 representatives of some 85 states signed four conventions which codified the existing maritime international law and included some new law. Although the Conference failed to reach agreement on the breadth of the territorial sea, it has generally been regarded as the most successful diplomatic conference ever held. All of the conventions have been ratified by the required number of signatory states, including the USA, Britain and the USSR, and are in force between the states having ratified them.

The Convention on the Territorial Sea and the Contiguous Zone establishes

---

*The views expressed in this paper do not necessarily represent those of the Department of External Affairs or the Department of National Defence.

rules for determining the territorial sea (but not its breadth), such as the application of the straight baseline system to deeply indented coasts and those with fringes of islands in their immediate vicinity, and of a 24-mile closing line for bays. It provides for a contiguous zone in which coastal states may exercise controls for customs, fiscal, immigration and sanitary purposes. However, no mention is made of an exclusive right of fishing in a contiguous zone since this question is tied to that of the breadth of the territorial sea on which no agreement could be reached. The convention also establishes principles governing navigation in the territorial sea, such as the right of innocent passage, and determines the rights applicable to foreign ships.

The Convention on the High Seas confirms the principle of the freedom of the high seas, including specifically the freedoms of navigation, fishing, laying submarine cables, and flying over the high seas. It also establishes general rules on the nationality of ships, jurisdiction, the right of hot pursuit, and the prevention of pollution of the high seas.

The Convention on Fishing and Conservation of the Living Resources of the High Seas specifies the conditions under which nations may exercise the right of fishing on the high seas, determines measures for the conservation of living resources, and establishes a procedure for the settlement of disputes over high seas fishing.

The Convention on the Continental Shelf defines the legal continental shelf as the seabed and subsoil of the submarine areas adjacent to the coast but outside the areas of the territorial sea to a depth of 200 metres or, beyond that limit, to where the depth of the superjacent waters admits of the exploitation of the natural resources of these areas, and the seabed and subsoil of similar submarine areas adjacent to the coasts of islands. The coastal state is given sovereign rights over the continental shelf for the purpose of exploring it and exploiting its natural resources.

The Territorial Sea

In theory a coastal state could claim all the waters up to half the distance from its coastline to the coastline of a state on the other side of an ocean. Legally, one must suppose, such a claim would be untenable.

The principal maritime nations have firmly opposed any proposal to extend territorial waters beyond 3 miles on the ground that it would adversely affect the freedom of the seas for commerce, naval operations and air traffic.

For reasons to be discussed shortly the USA and the USSR have each proposed in the UN General Assembly a draft convention to establish a 12-mile territorial sea. What would be the consequence of such a proposal?

A general extension of the territorial sea from 3 miles to 12 would reduce the area of the high seas by 3,000,000 square nautical miles. The same action would change the status of 116 straits from *international* or *high seas* straits to *territorial* straits, under the control of coastal states though always subject to the right of innocent passage. This general extension would create problems for national security and commercial interests for all nations, both at sea and in the air space above the territorial sea.

The Territorial Sea and Contiguous Zone Convention provides in Article 15(1) that "The coastal State must not hamper innocent passage through the territorial sea." Despite this provision, in current practice the right of innocent passage is not an absolute right. Thus the change of status of waters from high seas to territorial seas can have serious implications for both naval and commercial shipping. Also, there is no right of innocent passage for civil or military aircraft.

The law of comparative economic advantage can be severely upset by increased shipping costs resulting from the disruption or lengthening of international shipping routes. Ocean insurance rates fluctuate in response to political factors; the costs are passed on to the shipper and are paid ultimately by the buyer of goods.

The Contiguous Zone

In addition to exercising sovereignty in its territorial sea, a state may, for limited purposes, extend the national control into a part of the high seas known as "the contiguous zone." The zone may not extend beyond 12 miles from the baseline from which the territorial sea is measured.

Fishing Zones

The Territorial Sea and Contiguous Zone Convention includes no reference to a contiguous or other zone for exclusive fishing by the coastal state. However, Canada in 1964, like a number of other nations including Britain and the USA (in 1966), declared a 9-mile fishing zone contiguous to its territorial sea. In 1970 this contiguous fishing zone of Canada was incorporated in the territorial sea when it was extended to 12 miles from 3.

Fishing in the former 3-mile territorial sea of Canada is reserved exclusively for Canadians. An exception is provided in the case of the new fishing zones in the Gulf of St. Lawrence, Bay of Fundy, Dixon Entrance, Hecate Strait, and Queen Charlotte Sound and that part of the territorial sea that was formerly a fishing zone, (1) for states that have treaties with Canada (the USA and France) to continue fishing in the areas covered by the treaties, subject to re-negotiation, and (2) for those that have no treaties but which have a long or traditional practice, recognized by Canada, of fishing in certain areas, and in certain cases for particular fish stocks, off the east coast (Britain, Denmark, Italy, Norway, Portugal and Spain) to continue fishing in those areas until the end of a phase-out period that is being negotiated bilaterally in each case.

The main reason for the 1964 legislation which declared straight baselines for the territorial sea and provided a 9-mile fishing zone contiguous to the territorial sea was to protect the livelihood of Canadian fishermen by excluding foreign fishermen from the coastal areas. Applying the rule on straight baselines in the Territorial Sea Convention, the government has now promulgated straight baselines for the coasts of Labrador, the east and south coasts of Newfoundland, the south coast of Nova Scotia and the British Columbia coast. Most of Canada's seacoast, with the notable exception of the Arctic mainland and archipelago, has now been demarcated by straight baselines.

Legislation assented to in June 1970 declared a 12-mile territorial sea, to be drawn from whatever baselines are applicable. In the areas mentioned the territorial sea is measured from the straight baselines; in the other areas from the low-water mark along the coastline, following all the sinuosities. As there are no traditional fishing rights recognized in a state's territorial sea, the declaration of a 12-mile territorial sea will cancel, with the termination of the phase-out periods, all such rights now being enjoyed by certain countries between the 3- and 12-mile limits. Treaty rights, though, will continue to run, subject to possible modification.

The purposes of the 1970 amendment to the 1964 Territorial Sea and Fishing Zones Act was to separate exclusive fishing zones from the straight baselines drawn along our coasts; this permitted the establishment in March 1970 of "fisheries closing lines" across the entrances to the Gulf of St. Lawrence, Bay of Fundy, Dixon Entrance, Hecate Strait and Queen Charlotte Sound, an action which was strongly opposed by the USA and the distant-water fishing states affected.

It would be possible at a later stage to declare an exclusive fishing zone contiguous to the 12-mile territorial sea. As the world's best fishing grounds are in the waters above the continental shelf there has been pressure on the government to extend "Canadian fisheries waters" to the geological limit of the continental shelf. If this were done it would mean that to the east of Newfoundland, for example, Canada would be claiming an exclusive fishing zone of some 400 miles. Such a claim would exceed even that of Argentina, Brazil, Chile, Ecuador and Peru which claim exclusive fishing rights for their nationals to 12 miles but permit foreign fishermen to fish under licence in the 12- to 200-mile zone.

An additional reason for declaring a 12-mile territorial sea is that its adoption has made the Northwest Passage a territorial sea at the approximate mid-point, in the Barrow Strait where the islands are no more than 15 miles apart, giving Canada control over the entire passage. Prince of Wales Strait in the western Arctic was already territorial for part of its length because of a small island (Princess Royal) in the centre of the narrowest part of the strait. However, submarines can enter and leave Parry Channel through M'Clure Strait, under the ice, and for a short period about one year in nine the strait may be navigable by icebreaking surface vessels.

The High Seas

The expression "the high seas" refers to any part of the sea that is not included in the internal or inland waters or the territorial sea of any state. The high seas, it should be noted, are *not* the waters beyond the jurisdiction or control of any state because it is recognized in international law that coastal states have a right to exercise control for certain purposes in zones of the high seas contiguous to their territorial sea. Some coastal states have assumed additional rights, in some cases over very extensive areas of the high seas, but where these are claims to sovereignty it would seem they come into conflict with Article 2 of the High Seas Convention which states: "The high seas being open to all nations, no state may validly purport to subject any part of them to its sovereignty."

The underlying principle of the High Seas Convention is the assurance of maximum freedom of the seas. After identifying what freedoms are included in the freedom of the seas, Article 2 adds that "These freedoms, and others which are recognized by the general principles of international law, shall be exercised by all states with reasonable regard to the interests of other states in their exercise of the freedom of the high seas."

Consistent with the "freedom of the high seas" is the rule that, with very few exceptions, a vessel on the high seas is subject only to the jurisdiction of the country whose flag the vessel is authorized to fly; to this end every state, whether coastal or landlocked (the Swiss have an ocean-going merchant marine), has the right to sail ships under its flag on the high seas. Warships and ships owned or operated by governments on non-commercial service have, on the high seas, complete immunity from the jurisdiction of any state other than the flag state.

The Continental Shelf

Canada bases its offshore claims on the Continental Shelf Convention which defines the juridical shelf as extending from the outer limit of the territorial sea to where the water depth is 200 meters or, beyond that limit, to where the water depth permits exploitation of the underlying resources.

The convention recognizes that the coastal state "exercises over the continental shelf sovereign rights for the purpose of exploring it and exploiting its resources." Under the terms of the convention no one may explore the continental shelf or exploit its resources without the express consent of the coastal state. The rights of the coastal state do not depend on occupation or any express proclamation.

The convention establishes that the coastal state's rights do not affect the legal status of the waters above the continental shelf as high seas or that of the air space above these waters. Also it lays down the principles to be followed in determining the boundaries of the continental shelves of states whose coasts are opposite or adjacent.

The Continental Shelf Convention, which Canada regards as satisfactory, is being questioned, particularly by some less-developed countries; a trend seems to be developing in favour of restricting the limits of the continental shelf.

The exploitability test is so indefinite — in effect it permits a coastal state to exercise sovereign rights (assuming it has the technology to do so) to mid-ocean — that the world community will soon have to devise a more precise basis for determining the outer limit of national jurisdiction. Canada, which has a very wide shelf (at least on the Atlantic, where it extends some 400 miles to the east and south-east of Newfoundland), takes the position that the extent of national jurisdiction should be to the foot of the continental slope, that is comprising within the juridical shelf the entire geological shelf, not just the shallow part that slopes generally at less than a tenth of a degree (in the order of ten feet in a nautical mile).

States that have a narrow shelf or no shelf at all (including the landlocked states) generally hold a more restrictive view. They tend to favour various proposals for international schemes for trusteeship of the shelf areas beyond arbitrary limits of depth and/or distance. Canada, whose total shelf area in the Atlantic, Pacific and Arctic Oceans approximates 40% of the area of its land mass, would be penalized by either new criterion.

The Seabed and the Ocean Floor

Nations are now looking beyond the limits of the continental shelf. Recent studies in the UN have included peaceful uses of the seabed and a seabed arms control treaty.

International law is almost silent on the "peaceful" uses of the ocean floor. The rights of coastal states to the resources of the seabed beyond certain limits were not defined in the Law of the Sea Conventions. The view has developed, however, that any nation may explore the seabed and the subsoil of the ocean floor beyond the limits of its juridical or jurisdictional continental shelf and is entitled to retain any minerals it extracts.

The Seabed Arms Control Treaty was a separate issue. As it refers only to fixed objects and military installations on the seabed, its prohibitions do not extend to the operations (including bottoming or anchoring) of missile-armed submarines.

The Third Law of the Sea Conference

Preparations are now under way for a Third Law of the Sea Conference. The UN General Assembly's Ad Hoc Committee on Peaceful Uses of the Seabed Beyond the Limits of National Jurisdiction, of which Canada has been a member since its establishment in 1967, was expanded in membership from 42 to 86 and given a mandate to prepare for a comprehensive conference on the Law of the Sea, which is scheduled for 1973.

In the Preparatory Committee for the Law of the Sea Conference the USSR and Japan in particular have insisted that the conference should restrict itself to dealing with only three issues:
1. the breadth of the territorial sea,
2. passage through straits,
3. an international régime for the exploration and exploitation of the resources of the seabed beyond national jurisdiction.

Their position is in conflict with that of the majority of states in the committee, in particular the developing countries which are stressing the political nature of the outstanding Law of the Sea issues, including fisheries and marine pollution, and are making it clear that there are greater political complexities facing the Third Conference than existed in 1958 or 1960.

At the UN Law of the Sea Conference in 1958 and again at the Second Conference in 1960, as a compromise to prevent the adoption of more extensive territorial sea, Canada proposed (jointly with the USA in 1960) a "6-and-6" formula (a 6-mile territorial sea with a further 6-mile contiguous zone for certain purposes including exclusive fishing by the coastal state). Like all other proposals to the two earlier Law of the Sea Conferences it failed to be adopted, largely because of the continuous opposition of the Soviet bloc, states claiming a territorial sea beyond 6 miles, and others claiming very extensive fishing zones.

It became evident after the Second Geneva Conference in 1960, following the failure of first the 3-mile territorial sea (1930 at The Hague; 1958 and 1960), then the 6-mile territorial sea, that in order to ensure a two-thirds majority (required for the adoption of a substantive rule in the UN), any proposed solution must incorporate in some form the 12-mile territorial sea. (In 1958 over 80% of the states represented at Geneva voted in favour of a 12-mile territorial sea or contiguous fishing zone in one form or another.)

The USA and the USSR have jointly proposed a draft Convention on the Breadth of the Territorial Sea. The first article deals with the breadth of the territorial sea and proposes a 12-mile limit for the territorial sea and a contiguous fishing zone; the second deals with international straits and proposes that ships and aircraft should have unrestricted freedom of passage, as distinct from the right of innocent passage, through and over international straits affected by the adoption of a 12-mile limit; the third deals with fisheries conservation and the special interest of a coastal state in the high seas fisheries adjacent to its territorial sea, a proposal that is clearly intended as an inducement for coastal states such as Canada to accept the limitation of their jurisdiction at 12 miles.

Whereas a 12-mile breadth for the territorial sea now appears to be generally accepted as a rule of customary international law, the USA-USSR proposal is seeking to establish the 12-mile limit in conventional international law. The essential objective of the proposal is to stop at 12 miles the trend toward extensive claims to sovereignty over areas of the high seas, a trend that both sponsoring countries see as prejudicial to their strategic and fishing interests. The USA has alleged that the 12-mile limit represents the last opportunity for the world community to obtain agreement on a relatively narrow territorial sea.

The second article of the proposal is that in straits used for international navigation between one part of the high seas and another, or between the high seas and the territorial waters of a foreign state, all ships and aircraft should enjoy the same freedom of navigation and overflight for transit purposes as they do on the high seas. The proposal includes a provision that the coastal state may designate corridors or channels for such transit traffic. Also the proposal would not affect conventions or other international agreements already in force. The proposal on straits would affirm and extend the existing provision of international law for innocent passage which, in the view of the sponsoring countries, has proven to be inadequate, particularly in its application to warships, including submarines, and its omission of transit rights for aircraft through the superjacent air space.

The Right of Innocent Passage

Subject to the rules of the Territorial Sea Convention ships of all states have a right of innocent passage through its territorial sea. The existence of this right of innocent passage is the only limitation on the sovereignty of a coastal state in its territorial sea. It is a right that is normally granted to warships in peacetime; on this point the International Court of Justice held in the Corfu Channel (Merits) Case in 1949 that in time of peace warships have a right to pass through straits that are used for navigation between two parts of the high seas. The provision in the Territorial Sea Convention, which the USA and the USSR seek to amend, is that "there shall be no suspension of the innocent passage of foreign ships through straits which are used for international navigation between one part of the high seas and another part of the high seas or the territorial sea of a foreign state."

The USA, and presumably the USSR as well since its strategic requirements are similar, is not willing to rely on the usual right of innocent passage through straits. It is argued that the coastal state could deny the right of innocent passage and leave the foreign state seeking to exercise that right only the alternatives of accepting the denial or resorting to force or seeking an action before an international tribunal. Whereas the USSR has already (in 1918) adopted the 12-mile limit, the USA has, since 1793, opposed it or any extension beyond 3 miles and will now accept the 12-mile breadth for the territorial sea only if it is linked with a guarantee of free navigation in and above international straits, as distinct from a right of innocent passage, for vessels only, through such straits.

The USA is obviously very concerned at the increasing trend of coastal states to claim sovereignty or jurisdiction over areas of the high seas (the Americans call this "creeping jurisdiction") with its consequential encroachment on the freedom of the seas which the USA regards as essential to western defence interests.

Freedom of the Seas

"Freedom of the seas" is an interest shared by all nations. Because there is no international enforcement agency to ensure the protection of this common interest, the maritime states exercise almost complete jurisdiction over their shipping; this in itself is one aspect of freedom of the seas. But even the coastal states have some competence in the matter of jurisdiction, not only in their own waters but also on the high seas. Although some coastal states maintain naval forces for local protection only, most states with maritime interests maintain a navy which is charged with the protection of all those interests. For Canada one of the most important of these is the free movement of trade.

Economic Factors

Canada is the world's fifth largest international trading nation; trade is of vital importance to the Canadian economy. Canada has more of an interest than most countries in the unhindered flow of international commerce.

Almost all of Canada's overseas trade is carried in foreign-flag ships; the entire Canadian deep-sea merchant fleet carries only some 3% of the cargoes moving to and from Canadian ports. This almost complete dependence on foreign-flag carriers gives Canada a vested interest in the ability of these carriers to navigate freely on the world's oceans. Any action that limits the rights of nations to the free passage of their vessels on international trading routes affects Canada adversely by increasing the cost of shipping Canadian products to overseas markets.

Canada's policy has therefore been to support the general principle of the freedom of the seas, as it applies to all nations, and to rely on international law and the normal practice of other states to protect our trade in foreign-flag carriers. Although Canada has strongly supported the principle of the freedom of the seas in the past, in recent years, impelled by certain domestic factors, it has taken actions which are regarded, at least by some of its trading and military alliance partners, as contradictory to this principle.

Military Factors

Even for a country such as Canada that has no pretensions to being a major world power, maritime forces are a useful instrument of national policy. Warships are a traditional means of asserting and, if necessary, of sustaining a national presence. In addition to protecting the national interest in home waters, warships can be used abroad for purposes ranging from good-will visits and trade promotion to the emergency evacuation of citizens of the flag state from areas of political unrest. Naval vessels have been used on several occasions to transport military forces and their equipment to the theatre of peacekeeping operations.

In the context of defence it should be noted that of the eleven major ship-owning and -operating states (Britain, Denmark, France, Greece, Italy, Japan, the Netherlands, Norway, Sweden, the USA and the USSR) only Japan, Sweden and the USSR are not members of the Atlantic Alliance. The NATO nations together own and operate some 80% of the world's total shipping tonnage.

The evolution of the USSR as a first-rate seapower is one of the most important developments in the world power scene since the end of World War II. The Soviet Union has developed its seapower in all its aspects — naval, merchant marine, fishing, oceanographic research, ship construction and repair — to the extent that it is now the second most powerful maritime nation. This expansion has made possible a multi-ocean maritime strategy through which the Soviet naval capability, which is second only to that of the USA and still gaining, could be used to exert the influence and national aspirations of the USSR on a global basis.

The Interests of the Coastal States

Balanced against the free movement of shipping is the national interest of coastal states in their self-protection and the protection of their resources. This balance of interests (hopefully it will remain a balance and not become an international conflict over the uses and resources of the sea) is revealing itself in the meetings of the Preparatory Committee for the Third Law of the Sea Conference.

Though there is a general trend in support of a 12-mile territorial sea, the Latin-American group of states, recognizing that they are most unlikely to gain acceptance of a 200-mile territorial sea, are holding out for a 200-mile "economic zone" in which the coastal state would exercise jurisdiction (as some of them now do with fisheries) rather than sovereignty. They are receiving some support for their position, especially among the developing nations. On the other hand few countries seem to favour the USA-USSR proposal for freedom of navigation through international straits.

Of special importance to Canada is the question of fisheries conservation and jurisdiction for this purpose. Canada is working with other coastal waters fishing states (Iceland and some of the Latin-American countries in particular) in urging recognition that coastal states have special rights and interests in respect to the fisheries in waters adjacent to their territorial seas.

Whereas the USA–USSR proposal, which is supported by some of the major and more developed countries, is to limit the jurisdiction of coastal states to 12 miles and to restrict to a minimum any special rights or interests of those states beyond 12 miles, the trend is toward extended coastal state jurisdiction over fisheries, either on a basis of distance or zones, or of particular stocks of fish.

The USA introduced in the UN a draft Convention on the International Seabed Area; this is commonly referred to as the "Nixon Seabed Proposal." The claim by the USA in the Truman Proclamation of 28 September 1945 to the resources of its continental shelf led to the adoption at Geneva in 1958 of the Continental Shelf Convention which set a limiting depth at "200 metres or, beyond that limit, to where the depth of the superjacent waters admits of the exploitation of the natural resources."

The Nixon Proposal would now terminate national claims, except for a provision for baselines of less than 60 miles (a length that seems to favour US interests), at the 200-metre isobath. The outer shelf to the continental slope would become the International Seabed Area, a UN trusteeship zone to be administered by the adjacent coastal state. Revenues from permits and royalties would be shared by the world community, with preference being granted to the less developed states. So far the Proposal has received very little support.

The mandate given by the General Assembly to the Law of the Sea Conference covers the broad range of outstanding issues, including the question of marine scientific research on which there appears to be an increasing dichotomy of views between the developed and the developing countries. The latter, influenced by the fact that they lack sufficient resources and trained personnel to benefit from research in the oceans, are challenging the concept of unrestricted freedom of research advocated by some of the developed countries (those with advanced technology). They are taking a strong stand in favour of controls, whether by the coastal state or an international body, over marine scientific research. This is developing into one of the most controversial issues facing the Law of the Sea Conference.

Closely related is the issue of environmental protection in which Canada has undertaken a diplomatic initiative to develop international law for the prevention of marine pollution. Canada has proposed some legal principles for the protection of the marine environment and the prevention of pollution of the sea. These emphasize the priority of interests and responsibilities of coastal states to prevent marine pollution and define the right of a coastal state to extend its anti-pollution jurisdiction beyond its territorial waters, to intervene on the high seas in cases of shipping casualties that threaten to pollute the marine environment, and to obtain compensation for any damage suffered.

Although Canada received some support for its position, there is considerable resistance from the major shipping states (Britain, Japan, the USA and the

USSR) that see their interests threatened by any fundamental revision of the Law of the Sea.

Canada's Role

Canada has many friends and very few enemies in the world community. Since 1956, when preparations were in train for the first Law of the Sea Conference in 1958, Canada has been in partnership with the USA, Britain, and other maritime states in Law of the Sea matters. Now in spite of, and in some cases because of, its domestic actions on maritime claims, Canada enjoys a hard-won position of leadership in the development of the Law of Sea that is out of proportion to its actual status as a middle power. While Canada has taken some legislative steps that may be favourable to the nations opposing the great maritime powers, these actions are not sufficient in themselves to alter the fact that Canada's best national interests continue to lie *with* the maritime powers.

Conclusion

It is difficult to envisage the role of the developing countries — mostly, but not entirely, coastal states with few real maritime interests — in respect to the future Law of the Sea. Will these states make a positive contribution to the development of law? This scarcely seems likely. Will they block the efforts of the developed countries — mostly (like Canada) having both maritime and coastal interests — to develop the law? This seems the more likely role; the less-developed countries, it must be recognized, have enough votes to block *any* substantive proposal though probably they cannot muster the number needed to carry proposals of their own.

In the main the issues discussed in this paper represent the continuation of the historic struggle for control of the seas. If the achievement of a general accommodation on the outstanding issues of the Law of the Sea is the principal objective of the Third Conference, as Canada, the USA and some others see it, the years ahead will most certainly be difficult ones. Thus, while it is regretted that this paper is open-ended, you will appreciate that it cannot be otherwise.

9 From Polaris to the Future

IAN SMART
Assistant Director
International Institute for Strategic Studies

Not only because of its title but also because of its context, it seems entirely appropriate that this conference, which began by discussing *Dreadnought*, should end by considering *Polaris*. That span embraces a complete, coherent episode in the history of navies and of maritime strategy: from a capital ship invulnerable within its own marine environment to a capital ship which renders all things in another environment — the land — essentially and almost equally vulnerable. At one extreme, we have a naval instrument designed to dominate the sea: at the other, a vessel conceived solely to dominate the land.

At the beginning of this century, the principal role of naval forces was to match each other in combat. Only indirectly, by outweighing or overcoming its counterparts at sea, could a navy expect to affect military and political events on land. During the next fifty years, navies progressively brought their power to bear more directly on the areas of the globe above sea level — through amphibious action against coastal targets and, later, through action by carrier-based aircraft against targets far inland. Finally, in 1960, there appeared the apotheosis of that tendency, the nuclear-powered ballistic-missile submarine — dependent upon avoiding naval combat, incapable of conducting a naval battle with its primary armament and destined only to exert a direct influence over events on land. From the battleship, through the aircraft carrier, to the SSBN, the leading navies of the world have circumnavigated the intermediate obstacle of naval combat and have come directly ashore.

With the nuclear-powered ballistic-missile submarine — the SSBN — maritime strategy has extended its reach in another sense, into the world of strategic nuclear weapons and of deterrence. That extension has had a profound significance for both nuclear deterrence and navies.

By 1945, the 20th century development of amphibious warfare techniques and of naval air power had forced army commanders to take greater account of naval operations in their planning than at any time since the Napoleonic Wars. From the soldier's point of view, however, those operations, like operations in the air, remained essentially ancillary to land operations. National political leaders were, of course, constrained to attach progressively greater weight to both the sea and the air as military environments. In the one case, submarine warfare threatened to strangle their arteries of supply. In the other, strategic bombing struck directly at their industries and populations. Nevertheless, wars remained to be won or lost on land — as every war in at least the last 150 years has been. Military operations on land persisted as the touchstone of military strength — of the power to make war and to win it. Without the importance of naval power being denied, maritime strategy was taken, rightly or wrongly, to be a relatively discrete subject — a subject for the sailor, a subject, in some sense, for the specialist. The price of admiralty, in Kipling's phrase, might be high and the rewards of admiralty commensurately great, but the strategic implications of admiralty were still apparently finite within the military realm.

Nuclear weapons and the critical role of the SSBN in nuclear deterrence have changed all that. No discussion of war or peace between nuclear powers — of the possibility of war or the outcome of war — can be conducted, now or in the foreseeable future, without reference to the SSBN. No question of going to war can be considered, between such powers, primarily on the basis of the relationship between land-based forces, nuclear or conventional. Whether or not any such war can, in a traditional sense, be "won" must be doubtful. What is certain is that the decisions relevant to its onset and its potential outcome must all be taken, henceforward, in the shadow of the ballistic-missile submarine.

In such circumstances, maritime strategy can no longer be segregated from other aspects of strategic studies. The ballistic-missile submarine has not only brought the navy ashore; it has also brought the study of maritime strategy into the centre, and indeed the forefront, of strategic studies as a whole.

The say that the SSBN has carried maritime strategy into the age of deterrence is not to say that naval forces have had no previous deterrent role. But it is vital to distinguish between different forms of deterrence. In doing so, we shall find that the SSBN has, indeed, opened up a new area and added a new dimension to maritime strategy.

In some sense, all military forces and all military instruments throughout history can be said to have had a potential deterrent utility. Armed strength has always seemed to provide the means of deterring a would-be attacker. "Si vis pacem, para bellum." But the deterrent utility and threat which that axiom implies is of a particular type. It is, in fact, a threat of failure: a threat to defeat the effort of an adversary to achieve victory in military combat. The underlying assumption is that an enemy, seeing an insufficient chance of military success, will be deterred from launching an attack. Until the advent of nuclear weapons, the only deterrent task of military forces was, in general, to present just that kind of "failure" threat — to deter by demonstrating an ability and a will to defend successfully. *Par excellence*, that was, for example, the deterrent role of the British Navy at the beginning of this century, when *Dreadnought* was built: to deter attack on Britain by threatening to defeat any attempt to invade British territory or to interrupt British commerce.

The deterrent threat which strategic nuclear weapons convey is of a fundamentally different kind. It is a threat not of failure but of punishment: of punishment which will be incurred by the fact of an attack, rather than by its outcome, and which will be so heavy as to render the traditional rewards of successful attack insignificant, and even irrelevant. It is that version of deterrence — deterrence by threat of punishment — to which naval forces, in the shape, particularly, of the SSBN, now make a critical contribution.

To understand the importance of the SSBN in this context, we need to look back over the short history of nuclear deterrence itself. In 1945, immediately after Hiroshima and Nagasaki, many responsible observers were prepared to argue that the devastating power of nuclear armament would put an end to all war. Their fearful hopes have proved unfounded. Nuclear weapons have not eliminated war. They have not even prevented war between one nuclear power and the allies of another. In eastern, south-eastern and southern Asia, in Africa and in the Middle East, there have, since 1945, been wars which would have been considered major by any standards save those of the 20th century — and some whose scale even our haggard perspective has been unable to diminish. In some quarters, the consequent reaction against the exaggerated hopes and fears of 1945 has been proportionately strong. It has become almost fashionable to contend that nuclear weapons have deterred nothing, except possibly nuclear war between their immediate possessors, and that, in as far as they have done even that, they may only have forced violent conflict into other channels and other forms. The reaction is understandable. It may also be excessive. Despite the persistence, and even exacerbation, of their political conflicts since 1945, there has been no war of any sort between nuclear powers. There has been no clear case of a direct attack by any nation upon a nuclear power. There has been no attack of any kind upon a nation to whose defence with nuclear weapons a nuclear power has been formally, openly and generally committed. Against that background, it seems, at the least, rash to leap from a recognition that the operation of nuclear deterrence is obviously limited to a judgment that its utility is insignificant. The positive effectiveness of deterrence is essentially unprovable; its limits can only be located by its failures, and the use of the experimental method to that end is hardly likely to become popular. Nevertheless, it is arguably premature to dismiss the proposition that nuclear weapons do, indeed, deter the deliberate initiation of any attack, one plausible outcome of which would be their own use.

If that proposition is, in any way, reasonable, it can only be so subject to certain conditions which pertain to nuclear forces themselves. And it is exactly by enabling nuclear forces to satisfy those conditions that the maritime environment has now taken on such a special importance.

Again, the history of nuclear deterrence provides a key. Initially, nuclear weapons were designed to be dropped from aircraft, as they were in 1945. As between the United States and the Soviet Union, those aircraft were, in the main, based on land. Warned by radar of an enemy attack, they could take off and, once in the air, could represent an assurance of nuclear punishment which no adversary could reasonably ignore. As the 1950s wore on, however, the long-range ballistic missile loomed on the horizon, and the ability of land-based aircraft to survive a surprise attack seemed doomed by the prospect, to a fatal limitation. The "balance of terror," to use Winston Churchill's phrase, was becoming, in Albert Wohlstetter's, "delicate." It became more delicate still with the actual deployment of the first intercontinental ballistic missiles — ICBMs. These, handicapped by their liquid fuel systems and exposed on the earth's surface, were appallingly vulnerable to surprise attack by other ICBMs. Between such forces, in a time of acute crisis, the temptation to strike first might even seem irresistible to any leader who was prudent without necessarily being malicious. With deterrence dependent upon land-based missiles and aircraft which were mutually vulnerable to a surprise ICBM attack, neither government could have great confidence that it would, in fact, be able to deliver nuclear punishment. Nor could any government thus placed face with equanimity the prospect of reserving the use of its "punishment" forces until after the initiation of an enemy attack. Unless the second-strike forces required for deterrence by threat of punishment could be enabled to survive a surprise attack, nuclear weapons seemed doomed to destroy their own deterrent utility.

The solution was as obvious as the problem was urgent. The nuclear deterrent had to be rendered invulnerable to surprise attack by another force of its own kind. In both the United States and the Soviet Union, numerous courses were pursued in an effort to reach that end. Land-based aircraft were dispersed to more numerous airfields and an effort made to keep a proportion of them in flight at all times. Land-based missiles were redesigned so that they could be stored, invisibly and invulnerably, underground. Warning systems were improved and more widely deployed, so that second-strike forces would have a longer time in which to evade a surprise attack. In all cases, the objective was the same: to make deterrent weapons harder to find, harder to reach and harder to destroy.

The effort was substantial and expensive. In all the respects so far mentioned, however, it yielded nothing but partial and ephemeral solutions. Moreover, by a strange and striking irony, many of the attempts to make second-strike forces less vulnerable had the indirect effect of increasing their technical vulnerability. In order to enhance their chance of delivering relatiation, aircraft, for example, were designed or modified to fly at altitudes and speeds which would make them harder to detect and intercept. As a result, their technical ability to deliver a surprise first strike was also greatly improved. Land-based missiles were given — and continue to be given — greater accuracy and more effective warheads, with the result that they again became able to destroy other land-based missiles, even in reinforced underground silos. Space satellites and surface sensors were produced which could provide better information about a potential attacker and, as a result, produced exactly the information about the location and defence of land-based deterrent forces needed for a successful first strike against them. Although the effect may, for a time, have been muffled, nuclear weapons remained, on this evidence, their own worst enemies. The ICBM and the strategic bomber aircraft were still pitted directly against each other, in a situation of constant ambiguity concerning the intention of either side to strike first or second.

Only one result of the effort to make deterrent forces less vulnerable broke out of this spiral of directly interacting capabilities: the SSBN — the *Polaris* vessel with which this conference has chosen to end its proceedings. The nuclear-powered ballistic-missile submarine — over 80 examples of which are now in operation — has alone succeeded in combining a response to all three aspects of the threat of vulnerability which emerged in the late 1950s. It is harder to find, harder to reach and harder to destroy than any land-based deterrent system is now or can hope to be in the foreseeable future. It would be ridiculous to assume that its relative invulnerability to surprise attack will necessarily prove permanent. Neither technology nor strategy is accustomed to such permanence. And yet, despite the enormous sums spent on the development of anti-submarine warfare techniques, there is no apparent reason, at present or in a visible future, to question the ability of an SSBN force to survive any attack upon it to an extent which provides a substantial foundation for second-strike deterrence. The problems which face any nation seeking to destroy an opposing SSBN force are, after all, gargantuan. Destruction depends upon detecting, identifying, locating and tracking all, or at least the great majority, of the SSBNs concerned. It would, however, be hard to design an environment more unfriendly to a pursuer than the body of the ocean, with its differences of temperature and salinity and the intricate topography of its bed. In the long run, of course, the balance between the SSBN and its enemies may shift. For years to come, however, the dice which the ocean distributes to its users will be heavily loaded in favour of the hunted and against the hunter. That being so, there is no doubt that the SSBN has provided much the most effective solution so far found to the problem of "delicacy" in the strategic nuclear equation.

Even this is only one part – and perhaps the lesser part – of the contribution which the introduction of the SSBN has made to the effective operation of strategic deterrence. If the SSBN, as a response to the problem of vulnerability, had differed only in the degree of its effectiveness from other responses, such as the hardening of land-based missile silos or the dispersal of strategic aircraft, it would have been reasonable to expect its impact to be, like theirs, ephemeral: to expect that, like them, it would eventually be caught up again in the vicious spiral of directly interacting technical improvements. As it is, the SSBN shows every sign of being not only a quantitative but also a qualitative exception. For the first time, nuclear deterrent forces include, in the SSBN, a weapon system which is not, and will not become, its own worst enemy. The SSBN has done more than change one of the variables in the strategic equation; it has injected a new constant: a quantity which has the effect of biasing the equation itself in favour of second-strike deterrence and against first-strike attack.

Even against a force of land-based missiles and aircraft alone, the SSBN with its submarine-launched ballistic missiles – its SLBMs – offers only limited attractions as a first-strike weapon. For economic and technical reasons, the designer will want to install the largest reasonable number of missiles in each vessel. At the same time, he will be under strong pressure to keep its hull size to a minimum. The effect is to put a premium upon smaller, rather than larger, missiles. Given that premium, three factors enter into acute competition as far as those missiles are concerned: the desire for maximum range, the desire for maximum payload and the desire for maximum accuracy. Each costs both size and weight. In the circumstances, it is not surprising that SLBMs, in contrast to ICBMs, have generally lacked that combination of warhead size and accuracy which would enable them to attack land-based missiles, in their hardened, underground silos, with any substantial prospect of success, especially as the problem of achieving high accuracy is intrinsically greater in the case of a moving launch platform, such as a submarine, than in the case of a fixed, land-based launch silo. Nor, therefore, does the SLBM present any serious first-strike threat to the command and control systems of land-based forces, which are likely to offer targets even harder than ICBM silos. It has an undoubted ability to attack strategic warning radars but that, in relation to offensive forces, is entirely irrelevant, given that modern land-based missiles are designed to "ride out" an attack without having to be launched "on warning." Only to strategic bomber aircraft on their land bases or to the radar installations of an anti-ballistic missile – an ABM – system does the SLBM, in its present stage of development, offer any threat which might contribute to a first strike against a land-based deterrent.

The time may well come when, with technical improvement, the SLBM will have such accuracy that, even with a small warhead, it will be able to destroy an

ICBM silo or to attack land-based command and control systems effectively. At that point, it may, indeed, present a more credible threat to land-based deterrent forces. But at that point also, there will emerge all the more obviously what has always been the SLBM's — and the SSBN's — fundamental limitation as a first-strike weapon system: that it cannot threaten any other force of ballistic-missile submarines. The ICBM can threaten a first strike against other ICBMs, the bomber aircraft against other strategic bomber forces on the ground. Of all the strategic nuclear delivery systems so far developed, only the ballistic-missile submarine is not a natural cannibal which threatens to consume its own kind. As between two deterrent forces which consist only of land-based weapons, there is always at least a latent first-strike threat. As between deterrent forces which include substantial ballistic-missile submarine components, that ambiguity disappears. The invulnerability of the SSBN to its own kind, as well as to other deterrent weapons, not only relieves the SLBM of an implicit first-strike significance but also imposes a strong second-strike bias upon all other strategic nuclear weapons which are involved with it. Faced by a substantial force of ballistic-missile submarines, a first strike against land-based systems alone is obviously pointless. Equipped with a substantial force of such vessels, a pre-emptive or preventive strike at a moment of acute crisis is obviously unnecessary. It is in this way that the SSBN has injected a new constant into the equation which describes the relationship between nuclear deterrent forces.

The sole danger to the survival of the SSBN comes not from its own kind nor from other deterrent weapons but from the instruments of anti-submarine warfare — ASW. The United States and its NATO allies, in common with the Soviet Union, have been devoting great efforts to developing just those instruments to a point at which they may be able to challenge the SSBN effectively. So far, despite massive expenditure, their success has been minimal. Many nuclear-powered submarines can be detected and some can be identified, but all the resources of technology have so far been unable to track a signficant proportion of those submarines to an extent which would permit a successful and concerted attack to be made upon them. A great deal is heard at present about possible "break-throughs" in ASW. Many of the statements made are purely speculative. Most are apparently based upon a simple assumption that technology will inevitably, in time, produce a counter to any military weapon. Even if that assumption is generally justified, which is far from obvious, there is, however, no concrete indication in present technology of an efficient response to the SSBN.

Even if such a response became visible on the technical horizon, its desirability should, perhaps, be questioned. Between the world's two super-powers, divided in ideology and opposed in power, nuclear deterrence is at least *believed* to have been effective. That belief has, in itself, provided some considerable part of the confidence without which they would undoubtedly have been unwilling to engage in the political process of negotiation and accommodation upon which so much has come to depend. Without confidence in the effectiveness of deterrence, the first, tentative steps which we have recently been trying to take towards a genuine political *détente,* not only between the super-powers themselves but also between their allies, would have been unthinkable, just as the continuation of that process in the absence of adequate confidence would be impossible. Confidence in the effectiveness of deterrence between the United States and the Soviet Union has, in turn, come to depend critically upon the invulnerability of the ballistic-missile submarines which each has deployed. Yet each, assisted by its allies, persists in attempting to develop means of hunting down and destroying exactly those deterrent instruments upon whose demonstrable ability to survive its own political policies now depend for their success.

If ever there were a case for multilateral arms control in the interest of international security, it is thus, despite the obvious technical difficulties, in the area of ASW. Other justifications for ASW activity exist, but, in that neither convoy protection nor the more general safeguarding of surface shipping seem likely to be more than marginally relevant to any future strategic nuclear war, they are relatively insubstantial. Major powers with overseas commitments may, of course, feel that they will always need some anti-submarine capability for limited war contingencies. But the techniques relevant to that purpose are within the present "state of the art," and present no threat to the SSBN. It is the effort to counter the SSBN which still drives ahead the qualitative development of ASW forces, even though the success of that effort would seem likely to mark the fulfilment of a political and strategic death-wish on the part of its authors. A well-designed agreement to limit ASW development and activity would not only yield great financial savings but would also reinforce the structure of deterrence in a unique way. Even in the absence of an agreement, however, the attempt to combine active military efforts to defeat the ballistic-missile submarine with active political efforts to promote *détente* between East and West seems to be fraught with some obvious inconsistencies and dangers.

If I have concentrated upon the ballistic-missile submarine, I have done so, without apology, both because it marks one extreme of the conference's explicit scope and because I consider the SSBN to play a role of extraordinary strategic importance. I would not, however, wish to leave the impression that I see no significant future focus for maritime strategy beyond the operation of the SSBN and its potential enemies.

Within the context of future major war, the role of naval forces does, indeed, seem likely to be more limited than in the recent past. There has been a striking lack of combat between concentrated fleets in any of the limited wars which have taken place since 1945. Even in the Second World War, prolonged surface actions between large naval forces became exceptional. The extent to which air power might dominate large-scale naval combat and might, indeed, militate against any substantial concentration of battle squadrons was already apparent by 1945. Since that time, the further elaboration of air power and the introduction of the naval guided missile have encouraged that movement away from the traditional style of orchestrated fleet action.

The obsolescence of the battle fleet, the ambivalence of ASW and the marginal relevance of convoy protection in an unlimited war: each of these, in its own way, marks an actual or potential reduction in the utility of naval forces. But, even together, these tell only a small part of the story. Other naval roles, inherited from the past, must be sustained in the future. Some are likely to be of growing, rather than declining, importance. In addition, new roles must be assumed, and, in some cases, the ability and willingness to assume them may be critical not only to the proprietors of naval force but also to the security of the world at large.

Navies have possessed, and will retain, a unique potential for the application of political influence at long range and, above all, in a flexible manner. A naval "presence," sensitively demonstrated, will still carry significant political weight in carefully selected situations, whether far afield or around the borders of the nation state itself. The principal requirement, in this context, may be for naval units which are not so much individually sophisticated as they are collectively numerous. The role is one which calls for the greatest circumspection, in order that prudent "presence" may not be translated into imprudent "commitment," but it is not one whose importance seems likely to decline.

An extension of that role, which may be of growing significance in a world of nuclear weapons, is the ability of naval forces to contain certain local conflict situations and, by doing so, to prevent them from developing into limited or unlimited war. Naval units may be able to deter such a development, not by threatening punishment but by presenting threats of a different kind. In particular, by imposing a blockade or merely by interposing themselves within a conflict situation, naval forces may be able to threaten either failure or a more direct military involvement. The operations of the United States Navy during the Cuban missile crisis offer one example of the way in which naval vessels may simultaneously constitute both a barrier and a group of mobile hostages. The naval blockade of Rhodesia, the deterrent presence of Soviet naval vessels in Egyptian ports since 1967, the manoeuvre and reinforcement of the United States Sixth Fleet during the 1970 war in Jordan, the movement of Soviet and American squadrons towards the Bay of Bengal during the Indo-Pakistan war of 1971, all of these provide more recent examples of a naval technique which, although not always successful, may well become increasingly common.

Finally, there is one role which, while it is not entirely new, seems likely to take on a completely new importance during the next ten to twenty years and may, indeed, determine the future utility of some of the world's navies. Although it would still be premature to forecast a precise timetable, we are on the threshold of an explosive expansion in the exploitation of the deep ocean and its bed. National governments face the most serious problems in agreeing amongst themselves concerning the regulation of such activity. If they fail to agree, subnational and transnational groups, operating as independent actors on the international scene, may undertake that exploitation without the sanction or control of national governments – and, if they do, the result will be the creation of a new level of international anarchy, torn by its own conflicts, within which competing participants may eventually be driven to construct and use their own instruments of force. Even if national governments, facing that prospect, do agree upon the regulation of the deep ocean, their ability to avert anarchy will still, however, depend upon their ability to enforce such regulation and, by doing so, to bring "law and order" to the high seas. Without naval forces designed and trained for that purpose, prepared and equipped to understand and cope with the enormous problems of exploiting the ocean's resources, national governments will be powerless. That, indeed, may be the principal new challenge facing maritime strategy in the future. It is also a challenge which will bring together, in a special sense, the SSBN and the other instruments of naval power. A decade hence, the primary task for many of the world's navies may be to contain international conflict inside boundaries within which the SSBN can continue to make its critical contribution to the deterrence of major war. Failure in that task may permit the growth of a new maritime and international anarchy which, if it emerges, will make the deterrent power of the ballistic-missile submarine largely irrelevant.

References

1 An Historian and the Sublime Aspects of the Naval Profession

1. Mahan's Flag Officer on U.S.S. *Chicago,* who resented Mahan writing in his cabin in off-duty hours while at sea.
2. Cassell, 1965.
3. See, Gerald S. Graham. "The Ascendency of the Sailing Ship 1850-85," *Economic History Review,* vol. ix, no. 1, p. 75.
4. Lord Barham was First Lord in 1805.
5. 2 vols. (London, 1898).
6. It is not argued here that the Queen had the resources to do what subsequent historians like Corbett have argued she ought to have done.
7. (London, 1900).
8. He was first contacted in 1900.
9. Later Director of Naval Intelligence 1907-1909.
10. Naval historian and Admiral who wrote under the pen name 'Barfleur,' and who hated Sir John Fisher, the then First Sea Lord.
11. Source restricted.
12. Succeeded Fisher as First Sea Lord in 1910. The minute hereafter referred to was dated 1908.

13. Winston S. Churchill, *The World Crisis*. Vol. I (London, 1939), p. 69.

14. Viscount Oliver Fisher, *Journals and Letters of Reginald Viscount Esher*. Vol. 3 (London, 1938), p. 221.

15. Leonard Beaton was a distinguished Canadian defence journalist who worked in England. He died in 1971.

2 German Seapower

1. This legend's specificity made it more offensive than "In God We Trust," or a generalized maple leaf which disclaimed ancestry or provincial location. The *Britannica* approved the "properly designed tricolours" of France and Belgium, but found those of "the lesser American republics" confusing, and the "yellow, blue, red" of Venezuela "heraldically an abomination." H. Lawrence Swinburne, 11th ed., Vol. X, pp. 461-462.

2. Gordon Craig, *The Politics of the Prussian Army 1640-1945* (Oxford, 1955), p. 221. The most influential military planning body was the Prussian Army's General Staff, to which other staff officers were seconded. Walter Goerlitz does not see this as an Imperial General Staff, though one was possible under the German constitution. *History of the German General Staff 1657-1945*, trans., Brian Battershaw (New York, 1953), p. 94.

3. Craig, *Politics*, p. 222.

4. Berlin. See also H.U. Wehler, *Einleitung zu Ekart Kehr, Der Primat der Innenpolitik, Veröffentlichungen der Historischen Kommission zu Berlin* (Berlin, 1965), Vol. XIX. Steinberg, *Yesterday's Deterrent: Tirpitz and the Birth of the German Battle Fleet* (New York, 1965), p. 22, quoting Hubatsch, *Die Ara Tirpitz* (Gottingen, 1955), p. 83. Hubatsch's other major work is *Der Admiralstab und die obersten Marinebehörden in Deutschland, 1849-1945* (Frankfurt am Main, 1958). Marder's *The Anatomy of British Sea Power: A History of British Naval Policy in the Pre-Dreadnought Era, 1880-1905* (New York, 1949) was followed by his massive *From the Dreadnought to Scapa Flow: The Royal Navy in the Fisher Era, 1904-1919* (New York, 1961-1971). For the earlier German Navy see Frederic B.M. Hollyday, *Bismarck's Rival: A Political Biography of General and Admiral Albrecht von Stosch* (Durham, N.C., 1960).

5. *The Influence of Sea Power upon History 1660-1783* (Boston, 1890), pp. 23, 45.

6. *German Sea Power: Its Rise, Progress, and Economic Basis* (London, 1914), pp. 171, 373, 214, 285. Ashworth noted that 320 of the 46,471 students and 2486 of the 5653

Hörers were women, and that the post offices employed 319,026 men in 1907. Vol. XI, pp. 825, 823, 816. On the various German pressure groups see the review article by Otto Pflanze, "Another Crisis among German Historians? Helmut Böhme's *Deutschlands Weg zur Grossmacht," Journal of Modern History*, Vol. 40, No. 1 (March 1968), pp. 118-129, and Lamar Cecil, *Business and Politics in Imperial Germany, 1888-1918* (Princeton, 1967). On finance and the ways in which Tirpitz's naval policy led to desperate efforts to avoid struggles over taxation which might upset Reichstag approval for that policy, see Peter-Christian Witt, *Die Finanzpolitik des deutschen von 1903 bis 1913: Eine Studie zur Innenpolitik des Wilhelminischen Deutschland. Politische Studien, Heft* 415 (Lubeck, 1970). And for the ways in which German Chancellor Bernhard von Bulow torpedoed the Hague Conference of 1907 by claiming that it was a trick by the world's greatest naval power to freeze her supremacy when she had just completed a revolutionary type of battleship, see A.J.A. Morris, "The English Radicals' Campaign for Disarmament and the Hague Conference of 1907," *Journal of Modern History,* Vol. 43, No. 3 (Sept., 1971), pp. 367-393.

7. The phrase was one constantly used by the French *Jeune Ecole* and other partisans of commerce raiding in the 1880s. My "Continental Doctrines of Sea Power," and Margaret Tuttle Sprout's "Mahan: Evangelist of Sea Power" are in Edward Meade Earle, ed., *Makers of Modern Strategy: Military Thought from Machiavelli to Hitler* (Princeton, 1943), pp. 415-456.

8. London, 1925. *German Sea Power,* pp. 175, 192. In the *Britannica,* where the Royal Navy was dealt with under "Navy and Navies," though "United Kingdom" got both the British and Indian Armies, David Hannay saw the history of the German Navy as "one of foresight, calculation, consistency, and steady growth." He gave full credit to Tirpitz, "the first minister who was bred a seaman," and to General Stosch, but missed Admirals Alexander von Monts and Friedrich von Hollman. Vol. XIX, p. 310.

9. *German Sea Power,* p. 96. In 1886 the German Navy had only six flag officers budgeted. Two Admirals and one Vice-Admiral were members of the British Board of Admiralty, while two Admirals, six Vice-Admirals, and nine Rear-Admirals were on active duty. Lord Brassey, *The Naval Annual, 1886* (Portsmouth, 1886), pp. 400-401, 493. Other figures cited above are from later editions, or *Statistisches Jahrbuch fur die Deutsche Reich,* or *Historical Statistics of the United States,* or *The 1972 World Almanac and Book of Facts.*

10. *Influence,* p. 28.

11. *Compared and Contrasted with the Principles and Practice of Military Operations on Land* (Boston, 1911), p. 109.

12. *My Memoirs,* 2 vols. (Boston, 1919), Vol. I, p. 59. Hurd and Castle, *German Sea Power,* p. 348. Steinberg, *Yesterday's Deterrent,* p. 209.

13. *Studies in British Maritime Ascendancy* (Cambridge, 1965). Steinberg, *Yesterday's Deterrent,* p. 209.

14. *Naval Strategy,* pp. 108, 110, 105.

15. *Ibid.,* p. 16. Holger H. Herwig and David Trask, *Naval Operations Plans between Germany and the United States of America* (Munich, 1971) deal with some quite

theoretical studies. Some results of the confusion outlined in J.C.G. Röhl, *Germany without Bismarck: The Crisis of Government in the Second Reich, 1890-1900* (Berkeley, 1967) can be seen in Lothar Burchardt, *Friedenswirtschaft und Kriegsvorsorge: Deutschlands wirtschaftliche Rustungsbestregungen vor 1914.* Wehrwissenschaftliche Forschungen. Abteilung Militargeschichtliche Studien herausgegeben von Militargeschichtlichen Forschungsamt, No. 6 (Boppard am Rhein, 1968). Everyone wanted someone else to pay for planning, stockpiling, or civilian relief. Blockade was a military and thus a national matter, while relief was a responsibility of the states. Not that the British might not be just as confused. V.H. Rothwell, *British War Aims and Peace Diplomacy 1914-1918* (Oxford, 1971) shows that the Admiralty took the destruction of German naval power so much for granted that they never raised the issue before the end of the war. The British civilians were surprised, the French were shocked, and the Admiralty got its way.

16. *Struggle for the Sea* (London, 1959), p. 128. *My Life,* trans. Henry Drexel (Annapolis, 1960) is a different selection from *Mein Leben,* 2 vols. (Tubingen, 1956-1957). On the mutinies see Daniel Horn, *The German Naval Mutinies of World War I* (New Brunswick, N.J., 1969). Keith Bird, of the University of Bridgeport, is completing a major study of the Republican Navy.

17. Philip K. Lundeberg, "The German Naval Critique of the U-Boat Campaign, 1915-1918," *Military Affairs,* Vol. XXVII, No. 3 (Fall, 1963), pp. 117-118. Dr. Lundeberg is completing a major study of both U-Boat campaigns.

18. *Memoirs: Ten Years and Twenty Days,* trans. R.H. Stevens in collaboration with David Woodward (Cleveland, 1959), p. 153.

19. Steinberg, *Yesterday's Deterrent,* p. 211. Mahan, *Naval Strategy,* pp. 230-231.

20. These geographical facts were discovered by the Americans in 1941, when the Western Indian Ocean was made part of the Western Hemisphere's neutrality zone, well before Greece and Turkey were annexed to the North Atlantic. To return to the model suggested above, one might note (1) that Mahan's general ideas were more studied than his "qualifiers," and that he was less exclusively battleship minded than many Mahanites. But battleships were both practical and prestigious, and easiest to rate by quantifiable standards. (2) The selling of the High Seas Fleet by particular interest groups in (3) Germany's particular internal and international circumstances have also been well studied. (4) Marder has studied the British reply, but Germany's institutionalization of Mahan's ideas has not. One should not conclude, however, that this (5) made the later course of both wars inevitable. The success of the Schlieffen Plan or of Germany's improvised U-Boat war would have given the Anglo-Americans the very difficult task of retilting the balance against a powerful and entrenched army, an intact High Seas Fleet, and a submarine force which might have been much more effective in coastal defense than it was later. On the still more difficult problem of whether anyone yet has militarism or navalism quite right see Arno J. Mayer, "Internal Causes and Purposes of War in Europe, 1870-1956: A Research Assignment," *Journal of Modern History,* Vol. 41, No. 3 (Sept., 1969), pp. 291-303, and the comments by Peter Loewenberg, "Arno Mayer's 'Internal Causes and Purposes of War in Europe, 1870-1956' — an Inadequate Model of Human Behavior, National Conflict, and Historical Change," *Ibid.,* Vol. 42, No. 4 (Dec., 1970), pp. 628-636. On updating Mahan, one of the best and the most recent effort is James A. Barber, "Mahan and Naval Strategy in the Nuclear Age," *Naval War College Review,* Vol. XXIV, No. 7 (March 1972), pp. 78-88.

3 The R.N.A.S. in Combined Operations 1914-1915

1. Sir Walter Raleigh, *The War in the Air,* Vol. I, (London, 1922), pp. 143-145, 365.

2. See, Rear Admiral Murray F. Sueter, *Airmen and Noahs,* (London, 1928), *passim;* LORD FISHER'S MEMORANDUM ON AIRSHIPS AND ZEPPELINS, 1 May 1916, Arthur Marder, ed., *Fear God and Dread Nought, The Correspondence of Lord Fisher of Kilverstone,* Vol. III, (London, 1959), pp. 346-348; Churchill to his wife, 23 January 1916, Gilbert Martin, *Winston S. Churchill,* III, (London, 1971), p. 690; Report of a Meeting held in the First Sea Lord's Room on 3 August, 1915, S.W. Roskill, *Documents Relating to The Naval Air Service* (London, Navy Records Society, 1969), Doc. 75, p. 216; Balfour to Jellicoe, 26 August 1916, A. Temple Patterson, ed., *The Jellicoe Papers,* 1968), pp. 66-70; Air Cdre. C.R. Samson, *Fights and Flights* (London, 1931), p. 291; Arthur Marder, *From the Dreadnought to Scapa Flow,* Vol. II (London, 1965), p. 9; M.P.A. Hankey, *The Supreme Command, 1914-18,* Vol. I, (London, 1961), p. 75; Vice Admiral Sir Peter Gretton, *Former Naval Person* (London, 1968).

3. Cf. S.W. Roskill, *The Strategy of Sea Power* (London, 1962), pp. 136-8; J.M. Spaight, *The Beginnings of Organized Air Power* (London, 1927), which pays little attention to R.N.A.S. operations prior to 1916.

4. See, Hankey Memorandum of 12 June 1915, Roskill, *The Naval Air Service,* Doc. 70, pp. 210-11.

5. Pre-war concepts of naval aviation and early measures taken by the Admiralty may be traced in some of the following sources. *The Technical Index and History,* Vol. 40 (London, The Admiralty, 1920-21); Articles entitled "Air Power" and "The Effect of Aircraft on Naval Strategy," *Naval Review,* Vol. I, 1913, pp. 57-75, 256-269; H.C. Massey "The Seaplane and its Development," *R.U.S.I. Journal,* November 1913, p. 1452. Report of Captain of *Hermes* on use of aircraft during manoeuvres, 1913, Air Ministry Papers in the Public Record Office, London cited hereafter as Air 1/626/17/46; Raleigh, *War in the Air,* I, pp. 284-286, 357-362; Marder, *Dreadnought to Scapa Flow,* I, p. 353, II, pp. 4-6; C.F.S. Gamble, *The Story of a North Sea Air Station* (London, 1928), pp. 99-106; Air Chief Marshall Sir Arthur Longmore, *From Sea to Sky* (London, 1946); Randolph Churchill, *Winston S. Churchill,* Vol. II (London, 1967); Churchill to the Chief of Air Staff, 21 May 1919, Air 2/166/MR 17724/7, in which he describes taking up the cross channel steamers *Empress, Engadine* and *Riviera,* capable of 22.5 knots. Churchill claimed: "I ordered immediate preparations to be made for a bombing attack by torpedo seaplanes upon the German fleet in the roads at Wilhelmshaven." *Engadine* was converted, and her three seaplanes came near to attacking a German cruiser early in September. See also, *OU 6177, History of the Development of Torpedo Aircraft* (London, Admiralty 1919), pp. 1-5; Naval Staff Monograph (Historical) Vol. XI, *Home Waters,* Pt. II (London, Admiralty, December 1924), Appendix B, 202; Cf. J.S. Corbett, *Naval Operations,* I (London, 1920), p. 157. Murray F. Sueter, *The Evolution of the Submarine Boat, Mine and Torpedo* (Portsmouth, 1907) shows Sueter's approach to weapon development; Harald Penrose, *British Aviation,* Vol. I, *The Pioneer Years* (London, 1967), p. 346.

6. Churchill's memorandum of October, 1913, Raleigh, *War in the Air,* Vol. I, pp. 212-3, 263-276; C.F.S. Gamble, *The Air Weapon,* Vol. I (London, 1924), pp. 174, 239; See also Cd. 6067, "Memorandum on Military and Naval Aviation," Great Britain, *Parliamentary Papers,* 1912-13, Vol. li, pp. 567-578; The R.F.C. had been divided into naval and military wings on the naive assumption that there could be a purely land or naval war.

3

7. Vice-Admiral Richard Bell-Davies, *Sailor in the Air* (London, 1967), p. 94. This evidence is further supported by the Army Council's decision to send four aeroplanes to Eastchurch on 30 July so that Samson could go abroad if necessary. Raleigh, *War in the Air*, I, pp. 275-6; Churchill to Sir George Aston, 25 August 1914, Gilbert, *Churchill*, III, p. 56.

8. *Ibid.*; Samson, *Fights and Flights*, pp. 3-5; Churchill, *World Crisis*, Vol. I, p. 310; Naval Staff Monograph, *Home Waters*, Pt. II, Appendix A; Corbett, *Naval Operations*, Vol. I, p. 94.

9. Two passengers sat side-by-side in the front cockpit and the pilot behind in the rear cockpit. Owen Thetford, *British Naval Aircraft Since 1912* (London, 1962), p. 397.

10. Samson, *Fights and Flights*, pp. 3-5; Bruce Robertson, *British Military Aircraft Serials, 1912-1963* (London, 1964); Bell-Davies, *Sailor in the Air*, p. 95. As Bell-Davies says, this title had a somewhat seventeenth-century ring about it, and did not seem at all inappropriate under the circumstances.

11. Churchill to First Sea Lord, Fourth Sea Lord, Director of Air Department, 29 July 1914, *World Crisis*, I, p. 208.

12. Conclusions of an Admiralty Committee appointed in April, 1914, to investigate the question of camouflaging oil tanks and vulnerable points, H.A. Jones, *War in the Air*, Vol. III (London, 1931), pp. 78-9. Possibly the same document is cited as the Report of a Standing Sub-Committee of the C.I.D. in Marder, *Dreadnought to Scapa Flow*, I, p. 356.

13. *Sailor in the Air*, p. 96; cf. *Fights and Flights*, p. 14.

14. Naval Staff Monograph, *Home Waters*, Pt. I, p. 13n.; D.A.D. Memorandum, 1 September 1914, Air 1/671/17/128/1; Admiralty (Naval Attaché) to Marine Bordeaux, *ibid.*; Orders for Dunkerque, *ibid.*

15. Churchill to Sueter and Prince Louis, 5 September 1914, Churchill Papers, 8/67, cited in Gilbert, *Winston Churchill*, Vol. III, p. 67. See also Churchill to Board of Admiralty, *loc. cit.*; Samson to Admiralty, 4 September, 14 September 1914, Air 1/671/17/128/1. Asquith to Venetia Stanley, 19 September 1914, Montagu Papers, cited in Gilbert, *Churchill*, III, p. 74.

16. Gilbert, *Churchill*, III, p. 88; *Fights and Flights*, pp. 36-40, 54-89; *Sailor in the Air*, p. 96; Raleigh, *War in the Air*, I, pp. 387 ff.; Air 1/631/17/128/1.

17. Report from Squadron Commander Spenser Grey on the attack on Düsseldorf and Cologne, 17 October 1914, *ibid.*; Samson, *Fights and Flights*, p. 45.

18. *Ibid.*, p. 48.

19. Longmore, *From Sea to Sky*, p. 39.

20. They included an R.E.5, 120 H.P. Austro Daimler, a converted Sopwith seaplane, an 100 H.P. Anzani, a second converted Sopwith seaplane, two Bleriot monoplanes, an 80 H.P. Gnome, and a Sopwith 80 H.P. Gnome. Air 1/671/17/128/1; Robertson, *British Military Aircraft Serials*.

3

21. *From Sea to Sky,* p. 42.

22. *Sailor in the Air,* p. 106; Air 1/671/17/128/1; Robertson, *British Military Serials.*

23. *Fights and Flights,* p. 111.

24. *Sailor in the Air, loc. cit.;* Henderson to Haig, 8 September 1916 ". . . You had some experience of the Naval Air Service at the beginning of the war, and, take it all round, I do not think it has improved much since then," Air 1/2265/209/70/1.

25. Samson to Sueter, 6 December 1914, Air 1/671/17/128/2. In fact, as Churchill's correspondence shows, French indicated otherwise by 17 November. Churchill to French, 17 November 1914, cited in Gilbert, *Churchill,* III, pp. 162-3.

26. "I am too far in to go back" Churchill wrote to French, "and if when the force is completed you do not desire to have it with you, it can work with the Belgians." *Ibid.*

27. *Fights and Flights,* pp. 151-2.

28. Davies to Admiralty, 6 November 1914, "In accordance with my instructions from General Headquarters, I am returning to Dunkirk with my entire establishment." Air 1/671/17/128/2; Samson was furious, ". . . we had been kicked out of the war. . . ." *Sailor in the Air,* p. 110.

29. Orders issued by the Admiralty Air Department, 11 November 1914, Air 1/671/17/128/2.

30. They were used to locate the German troops advancing on Nieuport, 26 October to 7 November, during Rear Admiral Hood's bombardment; on 15 April 1915, after considerable experimentation and practice, seaplanes were first used to spot the fall of shot in Home Waters for H.M.S. *Excellent.* Naval Staff Monograph (Historical), Vol. VI, *The Dover Command,* Vol. 1 (London, The Admiralty, March 1922), pp. 14, 21. See also Commodore Sueter's paper on spotting, March 1915, Air 1/7/6/190.

31. *Sailor in the Air,* p. 113; *Fights and Flights,* p. 181; Air 1/681/17/128/2.

32. Samson to Sueter, 6 February 1915, cited in *Fights and Flights,* pp. 191-4.

33. The first night bombing raids were flown in December mainly for their nuisance value. *Fights and Flights,* p. 160.

34. *Ibid.,* pp. 194-202.

35. Jones, *War in the Air,* II, pp. 78-124, Appendix I; Colonel F.H. Sykes, Memorandum on R.F.C. Organization, 29 November 1914, Air 1/751/204/4/23; "A bombing scheme on an extended scale is being carried out by the French. 2 and 3 Wings are cooperating. . . . 1 Wing will send 10 machines. . . . These machines will be flown by pilots who can observe artillery fire singlehanded. Having dropped the bombs, the pilots will go on with artillery work." Holograph note in Sykes Memorandum of 29 November 1914, *ibid.*

36. Major W.D. Beatty to A.D.M.A., 25 April 1915, Air 1/141/15/40/308; Jones, *War in the Air,* II, Appendix I.

37. Penrose, *British Aviation*, I, p. 552; *War in the Air*, I, pp. 421, 471-5; R.N.A.S. Coastal Patrol Orders, 29 January-15 February 1915, Air 1/147/15/65.

38. Churchill to Secretary, 1st Sea Lord, 4th Sea Lord, D.A.D., 18 January 1915, *World Crisis*, II, pp. 539-40; Churchill to D.A.D., 3 April 1915, *ibid.*, p. 541; This minute is reproduced also in Roskill, *The Naval Air Service*, Doc. 65, p. 200. See also Minutes of a Conference held in the Admiralty on 3 April 1915, *ibid.*, Doc. 64, p. 195.

39. *Ibid.*

40. Wing Cdr. W.M.C. Moorsom, Report of Proceedings, H.M.S. *Ark Royal*, 5 February 1916, Air 1/649/17/122/418; Air Historical Board, interview with Lieut. Col. L.H. Strain (in 1915 Lieutenant, R.N.V.R., borne additional for *Ark Royal*), at the Bath Club, 30 May 1923, Air 1/726/137/5. Samson, *Fights and Flights*, pp. 207-9, 222-3; Report on the Performances of No. 1 Wing, R.N.A.S., during 1915; Roskill, *The Naval Air Service*, Doc. 93, p. 262, shows that Longmore was fully operational by 4 March at the latest. *Admiralty, Report of the Committee Appointed to Investigate the Attacks Delivered on and the Enemy Defences of the Dardanelles Straits, 1919* (London, April 1921) (Henceforth, *Dardanelles Report*), pp. 82-95.

41. Hankey memorandum to Dardanelles Commission, 1 February 1916, cited in Gilbert, *Churchill*, III, p. 291.

42. C.F. Aspinall-Oglander, *Gallipoli*, (London, 1929), p. 86n.; H.A. Jones, *War in the Air*, II, p. 24.

43. Churchill to Carden, 26 February 1916, *ibid.*, p. 304; Aspinall-Oglander, *Gallipoli*, I, p. 118.

44. Aspinall-Oglander, *Gallipoli*, p. 87.

45. Sir Ian Hamilton, *Gallipoli Diary*, Vol. I (London, 1920), pp. 8-9.

46. *Fights and Flights*, pp. 222-3.

47. *Gallipoli Diary*, I, pp. 110-11.

48. *Fights and Flights*, p. 224; *Dardanelles Report*, p. 518.

49. An observation also made of flights from Imbros. See E.L. Gerrard, personal reminiscences, Air 1/2301/212/7.

50. *Dardanelles Report*, p. 512.

51. *Ibid.*, p. 518; *Fights and Flights*, pp. 223-4.

52. *Gallipoli Diary*, p. 111. The Maurice Farmans photographed the peninsula up to a line drawn from Nibrunesi Pt. on the west, to the Maidos on the east, and on the Asiatic shore up to Chanak. When this was done a complete map was sent to the general staff at Lemnos. Major R.E.T. Hogg to War Office, 9 May 1915, Air 1/2119/207/72/2.

53. *Fights and Flights*, pp. 227, 229.

54. *Gallipoli Diary*, I, pp. 26, 113; *Dardanelles Report*, pp. 82-84.

55. *War in the Air,* II, p. 26; *Ark Royal* Report of Proceedings, 5 February 16, enclosure, "The Year's Work of H.M.S. 'Ark Royal' . . . " Air 1/649/17/122/418.

56. Account of R.N.A.S. Kite Balloon Development, Air 1/674/21/6/75; "Kite balloons for cooperation with the army and the navy" Air 1/1951/204/258/1; see also papers dated July, 1916, concerning the naval use of kite balloons, Air 2/127/B11683.

57. *Dardanelles Report,* p. 516.

58. *Manica* displaced 3500 tons, and had a speed of 11 knots.

59. R.N.A.S. Kite Balloon Development, Air 1/674/21/6/75.

60. *Gallipoli Diary,* pp. 109-113; Sykes report of 9 July 1915, Air 1/669/17/122/788.

61. "Naval Air Service – Reorganization," Admiralty Monthly Order 112/15, dated 1 March 1915; "Action against Aircraft," Confidential Admiralty Monthly Order 144/15, 1 July 1915, based on orders dated 27 February 1915; "R.N. Air Service," Admiralty Monthly Order 542/15, dated 1 October 1915, with effect from 1 August 1915.

62. *Gallipoli Diary,* pp. 112-3.

63. *Ibid.; Fights and Flights,* pp. 227, 266.

64. Major Hogg to War Office, 9 May, 10 June, 25 June, Air 1/2119/207/72/2; Col. F.H. Sykes to Admiralty 9 July 1915, Air 1/669/17/122/788; Wing Cdr. Longmore's report on aircraft in France, July 1915, Air 1/1310/204/12/1; *Dardanelles Report,* p. 518; *Fights and Flights,* pp. 224, 266.

65. *Dardanelles Report,* pp. 515-6.

66. *Dardanelles Report* p. 517; *War in the Air,* II, p. 56.

67. *Dardanelles Report,* pp. 514-5. H.S. Kerby, personal reminiscences, Air 1/2386/288/11/10.

68. When on 22 July *Ben-my-Chree* was called upon to provide spotting aircraft for monitors firing from the Rabbit Island anchorage, the ship was not prepared and had to borrow spotting personnel from the aeroplane squadron. *Ben-my-Chree* Report of Proceedings, 7 August 1915, Air 1/665/17/122/714. Sueter later explained that he had sent *Ben-my-Chree* principally to experiment with Torpedo-launching seaplanes against the Turk. Sueter to Vaughan-Lee, 20 December 1916, Adm. 1/8643, Marder Papers, (Extracts from microfilm loaned to Directorate of History, C.F.H.Q., by Arthur Marder) M32; *Airmen or Noahs,* p. 51; Naval Staff Monograph, *Home Waters,* Pt. IV, p. 231.

69. Beatty to Jellicoe, 31 May 1915, Fisher to Jellicoe, 31 May 1915, *The Jellicoe Papers,* I, pp. 164-5.

70. *War in the Air,* II, p. 355; Minute on Churchill's Memorandum on the Air Department dated 10 June 1915, Roskill, *The Naval Air Service,* pp. 207-10; cf. Hankey Memorandum of 12 June 1915, *ibid.,* no. 70, pp. 210-11; Gilbert, *Churchill,* III, p. 501.

3

71. The origin of Sykes' appointment remains a mystery. His autobiography gives no clue, nor do documents in the Air 1 or Air 2 series. See, *From Many Angles,* p. 159, *War in the Air,* II, p. 57, for the known facts. He was given the R.N.A.S. rank of Wing Captain, with the seniority of a Captain, Royal Navy, and a commission as Colonel, Royal Marine Light Infantry.

72. Sykes report of 9 July, Air 1/669/17/122/788; *War in the Air,* pp. 57-74; See also, *From Many Angles* and *Fights and Flights, passim.*

73. *Dardanelles Report,* p. 206.

74. Statement on Artillery by Brigadier General Hugh Simpson-Baikie, *Gallipoli Diary,* II, Appendix I; Sueter on spotting, March 1915, Air 1/7/6/190.

75. *Dardanelles Report,* pp. 30-31, 73, 78, 115-6, 177, 503-10, 520; Notes on operations of squadron covering the 8th Corps at Helles, Air 1/2317/223/21/124; *From Many Angles,* p. 167.

76. *War in the Air,* II, pp. 35-36; *Gallipoli Diary,* 23 June 1915, I, p. 323.

77. *War in the Air,* II, p. 65; *History of the Development of Torpedo Aircraft* (London, Admiralty, 1919), p. 5.

78. *Dardanelles Report,* pp. 206 ff.

79. Sykes Report of 21 October 1915, Air 1/654/17/122/503.

80. Dispatch of 11 December 1915, Cmd. 371, *Dardanelles Commission, Second Report,* 4 December 1917 in Great Britain, *Parliamentary Papers,* Vol. XIII, Pt. I, pp. 715-905.

81. Cf. Kemp, *Fleet Air Arm,* p. 47.

82. Aircraft Requirements in the Eastern Mediterranean, January 1916, Air 1/649/17/122/422.

83. Paper dated 19 November 1915, attached to Sykes report of 21 October 1915, Air 1/654/17/122/503. R. Adm. Vaughan-Lee the D.A.S., and R. Adm. H.F. Oliver, evidently found less difficulty in classifying duties. Sykes' papers were forwarded by the Admiralty to Vice Admiral, Mediterranean, "as large demands, such as are contained therein, cannot be dealt with on these lines. . ." They were to be resubmitted in a more concise form through regular channels. *Ibid.*

84. Papers concerning the naval use of kit balloons, 15 July – 29 August 1916, Air 2/127/B11683.

85. Gamble, *Story of a North Sea Air Station,* p. 139; Air 1/659/17/122/615.

86. Memorandum by C.-in-C., the Nore, February 1916, Air 1/147/15/72; Vaughan-Lee to 3rd Sea Lord, 25 April 1916, Air 1/149/15/104.

87. Hyde-Thompson to Sueter, 24 May 1916, Air 1/147/15/72; "TORPEDO CARRYING SEAPLANES," prepared by F/Cdr. Milne, submitted to Commodore Sueter for Commodore Bartolomé, 29 April 1916, *ibid.,* Cf. Gamble, *North Sea Air Station,* pp. 174-5.

3

88. Air 1/629/17/120; Thetford, *British Naval Aircraft*, p. 262; Gamble, *North Sea Air Station*, pp. 216-8.

89. *Fights and Flights,* pp. 291-2.

90. Wing Captain W.L. Elder, History of 3 Naval Wing, Luxeuil, Air 1/114/15/39/45 and Air 1/2107/207/42; *Navy List,* 18 December 1915. The Canadian pilots involved were G.R.S. Fleming, E.C. Potter, A.B. Shearer, L.E. Smith, T.W. Webber, D.H. Whittier, G.K. Williams.

4 The Dardanelles Revisited

1. This paper is based on fresh thinking since I prepared Volume II of my *From the Dreadnought to Scapa Flow* (London, 1965), stimulated by materials that were either unknown or unavailable to me then, more especially (1) the massive and highly significant "Mitchell Report," in which the essential facts are lost in a mass of verbiage and a faulty layout (*Report of Committee Appointed to Investigate the Attacks Delivered on and the Enemy Defences of the Dardanelles Straits. 1919,* C.B. 1550, 10 Oct. 1919, but printed in Apr. 1921: Commodore F.H. Mitchell was President of the Committee); (2) the De Robeck MSS. (Churchill College, Cambridge), which, alas, proved disappointing (there is little new material of consequence *from* De Robeck); and (3) the unpublished memoirs of Group Captain H.A. Williamson (Churchill College), which will be cited without reference to title or pagination. Additionally, I have profited greatly from an extensive correspondence with Williamson and Captain L.A.K. Boswell, R.N. Williamson, a pioneer in naval aviation, was Second-in-Command and Senior Flying Officer in the seaplane carrier H.M.S. *Ark Royal* during the critical first weeks of the operation. Boswell, who served at the Dardanelles in 1915 and has made a careful study of the naval side, admittedly has "a bee in my bonnet on the subject" of the fast mine-sweeping force and the difference that it could have made. I completely absolve these genetlemen from all the deficiencies that remain in this paper despite their best efforts.

2. Mitchell Report, p. 78.

3. Granted, it could, nevertheless, be argued that for Vice-Admiral S.H. Carden, who commanded the squadron at the Dardanelles, the reconnaissance meant a great deal. It was required to safeguard the landing party from surprise attack, to know about ships like the old Turkish Battleship *Barbarossa* (which, firing from Maidos, made the *Queen Elizabeth* move off on 6 March), and to know what the Turkish destroyers (always a possible menace to Allied ships) were doing.

4. Range spotting, as mentioned, could be done either by a ship in the Straits and in sight of the forts, or from the air. When Williamson went up to spot for the *Queen Elizabeth,* he received no instructions whatever. He was told nothing about a ship in

the Straits — direct spotting had been arranged from ships inside the Dardanelles — spotting for range, and so presumably was expected to give the necessary corrections for both range and direction, which he was quite prepared to do.

5. Mitchell Report, p. 425.

6. *Ibid.*, p. 49.

7. *Ibid.*, p. 484. It is unfortunate that paravanes were not yet available. The first set was fitted to a destroyer for trials in May 1915, and by August it was evident that this ingenious device for cutting mine moorings gave ships effective protection against mines. Marder, *From the Dreadnought to Scapa Flow,* II, 73-74.

8. Mitchell Report, p. 71. What follows on Turkish opinion is from *ibid.*, pp. 71, 432-33, as gleaned from information provided by the Turkish War Office and from interviews with German and Turkish officers.

9. *Proceedings of the Dardanelles Commission,* p. 173, Cab. 19/33, also Adm. 116/1437B (Public Record Office). And see *The Naval Memoirs of Admiral of the Fleet Sir Roger Keyes* (2 vols., London, 1934-35), i. 258-59, 266-69 (telegrams to the Admiralty of 22, 26, 27 Jan.) for De Robeck's reasons for not renewing the naval attack.

10. Winston S. Churchill, *The World Crisis* (6 vols., London, 1923-31), ii. 254, Churchill to Vice-Admiral E.P.A. Guépratte (commander of the French naval force at the Dardanelles), 9 Oct. 1930, Keyes MSS., 15/5 (Churchill College, Cambridge).

11. De Robeck's memorandum, "Orders for Combined Operations," 12 Apr. 1915, De Robeck MSS., 4/3

12. Mitchell Report, p. 179.

13. The air-spotting resources of the fleet were strengthened with the arrival of three aeroplane squadrons on 24 March with 18 aircraft. Only five were any good, but each of these could do three sorties a day, so that, even without the use of seaplanes, the *Queen Elizabeth* would have had continuous aircraft spotting. An aerodrome was established on Tenedos. H.M.S. *Monica,* the first kite balloon ship, arrived at Mudros on 9 April. At about 1,500 feet, the maximum height attainable by the kite balloon, observation was considered good. During April the balloon conducted "some most successful spotting operations" with the *Queen Elizabeth* and other battleships. So successful was the balloon that, initially, "its co-operation would seem to have been preferred to that of an aeroplane." Mitchell Report, p. 509.

14. The Turkish Fleet was organized into an Active Squadron (the ex-German battle cruiser *Goeben* and light-cruiser *Breslau,* and three old cruisers) under the German Admiral Souchon, and a Reserve Squadron (an ancient battleship, vintage 1874, reconstructed in 1902, and two of 1891, an old cruiser, and 10 torpedo boats) under a Turkish captain.

15. Churchill at an early date knew what they wanted of a defeated Turkey: "the surrender of everything Turkish in Europe." The terms of the Armistice might well

4

  include "the surrender of fortress of Adrianople and military positions affecting the control of the Bosphorus, Dardanelles & Constantinople," Allied occupation of Turkey in Europe, and Bulgarian occupation of the Enos-Media line. "Remember Cple is only a means to an end — & the only end is the march of the Balkan states against the Central Powers." Churchill to Grey, 28 Feb. 1915, F.O. 800/88 (Public Record Office).

16. Henry Morgenthau, *Secrets of the Bosphorus* (London, 1918), p. 148. (The U.S. edition has the title, *Ambassador Morgenthau's Story.*)

17. Mitchell Report, p. 433, and, similarly, pp. 71, 72.

5 Smaller Navies and Disarmament

1. Vice-Admiral B.B. Schofield, *British Sea Power in the Twentieth Century* (London, 1967), p. 105.

2. Robin Higham, *Armed Forces in Peacetime* (London, 1962), p. 132; also by the same author, *The Military Intellectuals In Britain: 1918-1939* (Rutgers University Press, 1966), pp. 33-34.

3. Captain S.W. Roskill, *Naval Policy Between the Wars, I: The Period of Anglo-American Antagonism, 1919-1929* (London, 1968), p. 534.

4. The Reconstruction Committee established in early 1918 to study the naval aspects of winding down the war. Eighteen volumes of deliberations and reports, contained in Admiralty Files (Public Record Office, London), Adm. 116/1745-1762. The Post-War Questions Committee, appointed in June 1919 to examine "in the light of the experience of the War, the military uses and values of the different types of war vessels . . ." Adm. 1/8586.

5. Sir Julian Corbett, *History of the Great War. Naval Operations,* Vol. III (London, 1924).

6. Adm. 167/61, Memorandum for the Cabinet, Navy Estimates and Naval Policy, 13 Feb. 1920; Adm. 167/62, Beatty to First Lord, "Naval Policy and Construction," 8 July 1920. Also Adm. 116/1175, "Naval Policy and Construction," Memo for the Cabinet by First Lord, 22 Nov. 1920.

7. *Fifty Years in the Royal Navy* (London, 1919), p. 332.

8. Adm. 116/3610, "Report of the Sub-Committee of the C.I.D. . . . on the Question of the Capital Ship in the Royal Navy," C.I.D. Paper, N-11, 2 Mar. 1921. See also, *The Times,* 14 Dec. 1920, "The Navy: An Ill-conceived Inquiry."

9. Adm. 116/1775, "Naval Construction," Admiralty Memoranda for the Cabinet, 10, 14 Dec. 1920; "The Retention of the Capital Ship," 14 Dec. 1920, C.I.D. Paper N-5.

10. *The Times,* 3 Jan. 1921.

11. Richmond Papers (National Maritime Museum, Greenwich), Diary, RIC/1/15, 10 Nov. 1920.

12. Adm. 116/3610.

13. *Ibid.*

14. *Ibid.,* Draft Report (Note by Mr. Churchill), dated 13 Feb. 1921.

15. Richmond Papers, RIC/7/4, Hankey to Richmond, 20 Dec. 1920.

16. Adm. 116/3445, C.I.D., "The Washington Conference," Note by the First Lord, 5 Oct. 1921, together with Memoranda by the Naval Staff.

17. M. Sullivan, *The Great Adventure at Washington* (London, 1922), p. xi.

18. Richmond Papers, RIC/7/2/3, Haldane to Richmond, 4 Dec. 1921.

19. Adm. 116/3445, C.I.D. 291-B, "Limitation of Tonnage of Battleships," Memo received from the Prime Minister, 1 Dec. 1921.

20. Adm. 116/3445, C.I.D. 297-B, Admiralty Memo, 2 Dec. 1921.

21. Richmond Papers, RIC/7/2/3, Field to Richmond, 7 May 1926.

22. *Ibid.,* RIC/7/1, Richmond to Henderson, 4 June 1929.

23. *Ibid.,* RIC/7/2/3, Richmond to Rear-Admiral Roger Bellairs, 17 Aug. 1929.

24. Admiral of the Fleet Sir Roger Backhouse (1878-1939): Third Sea Lord, 1928-32; C. in C. Home Fleet, 1935-38; First Sea Lord, 1938-39.

25. Richmond Papers, RIC/7/2/3, Richmond to Backhouse, 24 Aug. 1929.

26. *Ibid.,* RIC/7/4, Captain Bernard Acworth to Richmond, 16 Aug. 1929.

27. *Ibid.,* Acworth to Richmond, 20 Aug. 1929.

28. *Ibid.,* RIC/7/1, Richmond to Henderson, 5 Oct. 1929.

29. *The Times,* 21 Nov. 1929.

30. *Ibid.,* 23 Nov. 1929.

31. Richmond Papers, RIC/7/2/3, Sir Oswyn Murray to Richmond, 28 Nov. 1929; Richmond to Murray, 6 Dec. 1929.

5

32. 14 Dec. 1929.

33. Jan. 1930, pp. 251-254.

34. Admiral of the Fleet Lord Chatfield, *It Might Happen Again* (London, 1947), p. 60.

35. Richmond Papers, RIC/6/3, "The Naval Conference, I and II," letters to *The Times*, undated (1930). (Not Published).

36. *Ibid.*, RIC/7/2/2, Cabinet Committee on Bombs and Battleships, 1936. Richmond's correspondence and memoranda submitted to the C.I.D. and Admiralty.

37. *The Liddell Hart Memoirs* (London, 1965), I, pp. 284-285.

38. *Ibid.*, p. 186.

39. In conversation with the author at States House, Medmenham, 3 July 1969.

40. See, Liddell Hart Papers, Richmond to Liddell Hart, 2 Oct. 24 Nov. 1930, and 20 Jan. 1931.

6 Canadian Maritime Strategy in the Seventies

1. 60 of the 70 RCN frigates and 106 of the 122 corvettes were built in Canada.

2. The Canadian carrier had fighter as well as fixed-wing antisubmarine aircraft.

3. Canadian Naval Operations in Korean Waters, 1950-55. T. Thorgrimsson and E.C. Russell. Dept. of National Defence, Ottawa (1965).

4. Defence in the 70s. White Paper on Defence. Information Canada. August 1971.

5. Tenth Report of the Standing Committee on External Affairs and National Defence Respecting Maritime Forces. Queen's Printer. 1970.

1 2 3 4 5 — 77 76 75 74 73

SOUTHEASTERN MASSACHUSETTS UNIVERSITY
V163.C66 1972
Dreadnought to Polaris: maritime strateg

3 2922 00191 880 1